New Developments in Medical Research

Antimicrobial Potential of Essential Oils

NEW DEVELOPMENTS IN MEDICAL RESEARCH

Additional books and e-books in this series can be found on Nova's website under the Series tab.

New Developments in Medical Research

Antimicrobial Potential of Essential Oils

Bruno Oliveira de Veras
Yago Queiroz dos Santos
Fernanda Granja da Silva Oliveira
Marcia Vanusa da Silva
Jackson Roberto Guedes da Silva
and
Maria Tereza dos Santos Correia
Editors

Copyright © 2020 by Nova Science Publishers, Inc.

All rights reserved. No part of this book may be reproduced, stored in a retrieval system or transmitted in any form or by any means: electronic, electrostatic, magnetic, tape, mechanical photocopying, recording or otherwise without the written permission of the Publisher.

We have partnered with Copyright Clearance Center to make it easy for you to obtain permissions to reuse content from this publication. Simply navigate to this publication's page on Nova's website and locate the "Get Permission" button below the title description. This button is linked directly to the title's permission page on copyright.com. Alternatively, you can visit copyright.com and search by title, ISBN, or ISSN.

For further questions about using the service on copyright.com, please contact:
Copyright Clearance Center
Phone: +1-(978) 750-8400   Fax: +1-(978) 750-4470   E-mail: info@copyright.com.

NOTICE TO THE READER

The Publisher has taken reasonable care in the preparation of this book, but makes no expressed or implied warranty of any kind and assumes no responsibility for any errors or omissions. No liability is assumed for incidental or consequential damages in connection with or arising out of information contained in this book. The Publisher shall not be liable for any special, consequential, or **exemplary damages resulting, in whole or in part, from the readers' use of,** or reliance upon, this material. Any parts of this book based on government reports are so indicated and copyright is claimed for those parts to the extent applicable to compilations of such works.

Independent verification should be sought for any data, advice or recommendations contained in this book. In addition, no responsibility is assumed by the Publisher for any injury and/or damage to persons or property arising from any methods, products, instructions, ideas or otherwise contained in this publication.

This publication is designed to provide accurate and authoritative information with regard to the subject matter covered herein. It is sold with the clear understanding that the Publisher is not engaged in rendering legal or any other professional services. If legal or any other expert assistance is required, the services of a competent person should be sought. FROM A DECLARATION OF PARTICIPANTS JOINTLY ADOPTED BY A COMMITTEE OF THE AMERICAN BAR ASSOCIATION AND A COMMITTEE OF PUBLISHERS.

Additional color graphics may be available in the e-book version of this book.

Library of Congress Cataloging-in-Publication Data

Names: Oliveira de Veras, Bruno, editor. | Queiroz dos Santos, Yago, editor. | Vanusa da Silva, Marcia, editor. | Correia, Maria Tereza dos Santos, editor. | Granja da Silva Oliveira, Fernanda, editor. | Guedes da Silva, Jackson, editor.
Title: Antimicrobial potential of essential oils / [edited by Queiroz dos Santos Yago, Vanusa da Silva Marcia, Tereza dos Santos Correia Maria, Oliveira de Veras Bruno, Granja da Silva Oliveira Fernanda, and Guedes da Silva Jackson].
Description: New York : Nova Science Publishers, Inc., [2019] | Series: New developments in medical research | Includes bibliographical references and index. |
Identifiers: LCCN 2019056803 (print) | LCCN 2019056804 (ebook) | ISBN 9781536169454 (paperback) | ISBN 9781536169461 (adobe pdf)
Subjects: LCSH: Essences and essential oils--Therapeutic use. | Anti-infective agents.
Classification: LCC RM666.A68 A57 2019 (print) | LCC RM666.A68 (ebook) | DDC 615.7/92--dc23
LC record available at https://lccn.loc.gov/2019056803
LC ebook record available at https://lccn.loc.gov/2019056804

Published by Nova Science Publishers, Inc. † New York

Contents

Preface		**ix**
Chapter 1	Mechanisms of Action and Antimicrobial and Antibiofilm Activity of Essential Oils *Katharina Marques Diniz,* *Bruno Oliveira de Veras,* *Fernanda Luizy Aguiar da Silva,* *Carla Bismarck Lopes,* *Sarah Romini de Lima Basto* *and Thayza Christina Montenegro Stamford*	**1**
Chapter 2	Antibacterial Activity of Essential Oils in Aromatherapy Protocols *Yago Queiroz dos Santos,* *Gabriella Silva Campos Carelli,* *Anderson Felipe Jácome de França,* *Clecia de Carvalho Marques,* *Marcia Vanusa da Silva,* *Maria Betânia Melo de Oliveira,* *Ana Catarina de Souza Lopes* *and Bruno Oliveira de Veras*	**21**

Chapter 3	Mechanisms of Bacterial Resistance to Antibiotics: Essential Oils as a Strategic Tool	45

Jalcinês da Costa Pereira,
Jackelly Felipe de Oliveira,
Rádamis Barbosa Castor, Lucas Silva Brito,
Samuel de Souza Soares,
Maria Vanessa Pontes da Costa Espínola,
Fernanda Pereira Santos,
Bruno Oliveira de Veras
and Krystyna Gorlach-Lira

Chapter 4	Essential Oil of the Tea Tree (*Melaleuca alternifolia*): A Potent Antimicrobial	69

Krystyna Gorlach-Lira,
Maria Vanessa Pontes da Costa Espínola,
Lucas Silva Brito, Fernanda Pereira Santos,
Jackelly Felipe de Oliveira,
Samuel de Souza Soares,
Rádamis Barbosa Castor,
Jalcinês da Costa Pereira and
Bruno Veras de Oliveira

Chapter 5	Antimicrobial Activity of Essential Oils from Caatinga Plant Species	93

José Rafael da Silva Araujo,
Paulo Henrique Valença Nunes,
Camila Marinho da Silva, Suyane de Deus e Melo,
Marx Oliveira de Lima, Silvany de Sousa Araujo
and Bruno Oliveira de Veras

Chapter 6	Antibacterial Potential of Essential Oil from *Syzygium aromaticum* (L.) Merr. and L. M. Perry João Ricardhis Saturnino de Oliveira, Cristiane Marinho Uchôa Lopes, Rebeca Xavier da Cunha, Francisco Henrique da Silva, Bruno Oliveira de Veras, José Galberto Martins da Costa, Vera Cristina Oliveira de Carvalho and Vera Lúcia de Menezes Lima	**129**
Chapter 7	Antibacterial Activity of Essential Oils from Species of *Annona* L. João Ricardhis Saturnino de Oliveira, Weber Melo Nascimento, Ana Paula Sant'Anna da Silva, Vera Cristina Oliveira de Carvalho, Bruno Oliveira de Veras, Bianka Santana dos Santos and Vera Lúcia de Menezes Lima	**143**
Chapter 8	Antimicrobial Activity of Essential Oil from *Xylopia frutescens* Aubl (*Annonaceae*) Amanda Virginia Barbos, Bárbara de Azevedo Ramos, Milena Martins Correia da Silva, Claudio Augusto Gomes da Camara, Rafael Artur Cavalcanti Queiroz de Sá, Francisco Henrique da Silva, Maria Tereza dos Santos Correia and Marcia Vanusa da Silva	**155**

About the Editors — **173**

Index — **177**

Related Nova Publication — **181**

PREFACE

Instead of relying on prescription medications with numerous dangerous side effects, what if you could opt for a safer, natural alternative to address your health concerns? Medicinal plants for therapeutic purposes have been used for many years. The antimicrobial activity of essential oils and their major constituents has been widely documented by several works, however, in a fragmented way. Based on this premise, this book is designed to provide an overview of current knowledge about the antimicrobial properties of essential oils and their mechanisms of action, either alone or in combination, as a possible tool for obtaining new antibiotics.

Chapter 1 - The emergence of microorganisms with multiple resistance to antibiotics has prioritized the search for biocides that are biosustainable, safe to use, and have action in the modeling of microbial resistance. Essential oils, for the most part, demonstrate these principles required for new antimicrobials. They show inhibitory action against Gram-positive and Gram-negative bacteria, fungi, and some viruses. Some also have antiparasitic and insecticidal activity. The mechanisms of antimicrobial action include alteration of the cytoplasmic membrane, enzymatic inhibition, and even alterations in DNA, causing the death of the cell. Due to their instability under storage conditions, methods have been studied that allow transport and release of the oils while maintaining their stability and prolonging shelf life. The properties of essential oils have several applications in the pharmaceutical, cosmetics, food, and other industries.

Chapter 2 - Aromatherapy is the therapy that employs plantsvolatile aromatic elements, that is, their essential oils. The practice that helps patients to sleep and rest as well as helps on alertness, creativity, among others. The therapeutic massage integrates a set of practices with maneuvers whose goal is to promote health and balance with the body, promoting psychological effects on the skin, visceral pain. This pharmacological components of essential oils are volatile constituents at temperature environment, most of which originate from secondary metabolism produced and stored in their own secretory structures formed in the leaves, flowers, branches, stems or roots of various species usually secreted by glandular trichomes, which have various forms, structures and functions distributed mainly on the surface of the leaves. The identification of essential oil constituents is important for the understanding and prediction of their physiological effects where main studied activities are antimicrobial activity, namely antibacterial, antifungal and antiviral, anxiolytic, antioxidant, anticarcinogenic and antinociceptive consisting a helpful and potential pharmacological sources in order to develop new clinical aromatherapy protocols.

Chapter 3 - It is now well established that bacterial resistance to antibiotics has become a serious problem of public health that concerns almost all antibacterial agents and that manifests in all fields of their application. The increasing number of antibiotic resistant bacteria, through various mechanisms, such as efflux pumps and β-lactamase production, threatens the world due to the possible emergence of highly pathogenic bacteria resistant to treatments using conventional antibiotics. Based on this perspective, a large number of studies are currently being conducted in search of new alternative drugs and tools to combat resistant bacterial pathogens. Essential oils enter this scenario as an alternative option to antimicrobial resistance, as they present remarkable antimicrobial action acting exclusively or in combination with antibiotics, increasing the susceptibility of these pathogens to drugs currently in use. Based on this perspective, the present work was elaborated in a review focused on two mechanisms of bacterial resistance to antibiotics (efflux pump and β-lactamase production), and the possibility of applying essential oils alone and in combination of antibiotics to overcome these mechanisms.

Chapter 4 - Tea Tree (*Melaleuca alternifolia*) is a medicinal aromatic plant found in tropical and subtropical regions and known for its antimicrobial properties. The essential oil of Tea Tree (TTO) is a complex mixture of terpenes, mainly mono and sesquiterpenes, of which terpinen-4-ol is the most representative. A number of studies have shown the effectiveness of TTO in combating human and animal pathogenic microorganisms, and even those resistant to antibiotics such as methicillin-resistant *Staphylococcus aureus* (MRSA) and fluconazole-resistant *Candida albicans*. But despite widely investigated antimicrobial activity, the mechanisms of action by which TTO works remain uncertain. There are evidences that TTO, acts as a causative agent of disruption of the fungal and bacterial cell wall and plasma membrane, with consequent loss of metabolites and cell death. This revision focus on essential oil of *M. alternifolia* as the antibacterial and antifungal agent to be important alternative to conventional methods employed in the treatment of microbial infections, as well as brings insight to its mechanism of action against bacteria and fungi.

Chapter 5 - Caatinga is a unique Brazilian cultural formation and much of the biological heritage is not found anywhere in the world. This biome possesses a variety of plant species with the primary vegetation still preserved in some areas. Its species have physiological characteristics that reflect complex and peculiar adaptations to the exceptional conditions of this environment, which aroused the interest of the scientific community. This study focused on plant species that present essential oils (EOs) with antimicrobial potential and that occurs in the Caatinga Phytogeographical Domain (CPD). Three botanical families with occurrence in CPD were select for the current study: Euphorbiaceae, Asteraceae, and Lamiaceae. β-Caryophyllene was one of the most recurrent compounds, being responsible for this activity. Among the classes of microorganisms analyzed, all families showed activity against Gram-positive bacteria and some species of fungi. However, the Lamiaceae family, besides being effective against the mentioned classes, showed activity against Gram-negative bacteria as well. The expansion of biochemical prospection of Caatinga plants is extremely important since its plant species have been the source of biomolecules that can become new alternatives for antimicrobial drugs.

Chapter 6 - *Syzygium aromaticum* is used in culinary, but this herb is also used for anti-inflammatory and antimicrobial purposes. The aim of this study was to investigate reports of *S. aromaticum* essential oil antimicrobial activity. Literature review was conducted in several databases, and data from essential oil related to bacteria strains were collected. There are reports of *S. aromaticum* essential oil against 52 strains of bacteria; while Eugenol was tested against 33. S. aromaticum killed all strains tested, but had better results against *Vibrio choleare*. Thus, *S. aromaticum* essential oil might be used as a natural tool against bacteria.

Chapter 7 - *Annona* is one of the most important genus from *Annonaceae*. Several studies indicate activity and new biocompounds found in extracts from leaves, barks, and fruits from the genus, although little is known about essential oil possibilities. This study aimed to review antibacterial activity of species from the genus *Annona*. This review searched antibacterial studies from all Annona species, without excluding date of publication, language, and strains. Eight species of Annona were reported against several strains. *Bacillus* and *Staphylococcus* species were the most tested, and Annonas had the better results for these strains. Studies diverge method of antibacterial activity and strains, which turns difficult to compare essential oils capabilities. *A. squamosa* and *A. cherimola* were the most tested, and had the highest potential.

Chapter 8 - Plants with medicinal properties have been used by humans since ancient times as an alternative resource of treatment for various diseases. *Xylopia frutescens* popularly known as "embira", is used in folk medicine and produces essential oil from its secondary metabolism. The use of essential oils in phytotherapy is associated to various pharmacological actions and are extensively studied mainly for their antimicrobial activity. Thus, this study aims to chemically characterize the components present in the essential oil of leaves of *X. frutescens* and evaluation of its antimicrobial activity. Twenty five components have been identified in the essential oil. The minimal inhibitory concentration (MIC) of the essential oil was at 4 mg / mL for all tested strains of *Staphylococcus aureus*. In conclusion, this is the first report on analysis of volatile compounds and antimicrobial activity from *X. frutescens* essential oil of leaves extracted from Atlantic Forest in the Pernambuco, Brazil. The essential oil of *X. frutescens* showed satisfactory MIC results for all *S. aureus* strains, indicating a promising source of biomolecules for development of new antibacterial drugs.

In: Antimicrobial Potential of Essential Oils ISBN: 978-1-53616-945-4
Editors: B. Oliveira de Veras et al. © 2020 Nova Science Publishers, Inc.

Chapter 1

MECHANISMS OF ACTION AND ANTIMICROBIAL AND ANTIBIOFILM ACTIVITY OF ESSENTIAL OILS

Katharina Marques Diniz, Bruno Oliveira de Veras, Fernanda Luizy Aguiar da Silva, Carla Bismarck Lopes, Sarah Romini de Lima Basto
and Thayza Christina Montenegro Stamford
Department of Tropical Medicine, Federal University of Pernambuco,
Recife, Pernambuco, Brazil

ABSTRACT

The emergence of microorganisms with multiple resistance to antibiotics has prioritized the search for biocides that are biosustainable, safe to use, and have action in the modeling of microbial resistance. Essential oils, for the most part, demonstrate these principles required for new antimicrobials. They show inhibitory action against Gram-positive and Gram-negative bacteria, fungi, and some viruses. Some also have antiparasitic and insecticidal activity. The mechanisms of antimicrobial action include alteration of the cytoplasmic membrane, enzymatic inhibition, and even alterations in DNA, causing the death of the cell. Due to their instability under storage conditions, methods have been studied that allow transport and release of the oils while maintaining their stability and prolonging shelf life. The properties of essential oils have several applications in the pharmaceutical, cosmetics, food, and other industries.

INTRODUCTION

Medicinal plants and their compounds, such as vegetable oils and essential oils, have been used since the beginning of humanity as therapeutic agents for various pathological conditions. Scientific research has contributed to the elucidation of the therapeutic properties of these compounds (Auddy et al. 2003; Costa et al. 2015; Salvia-Trujillo et al. 2015; Luz et al. 2018).

Among plant compounds, essential oils have received special attention. Essential oil concentration in different plant parts varies quantitatively and qualitatively according to several factors, such as soil type, soil fertility, climate, period of the day, time of the year (seasonality), etc. These secondary metabolites in aromatic plants are important defenses against animal and insect attacks as well as being attractive agents for some pollinating insects (Dudareva and Pichersky 2008; Ali et al. 2015).

The chemical composition of an essential oil is determined by genetic factors; however, other aspects can promote considerable variation in the production of secondary metabolites. In fact, these metabolites can be produced as a result of chemical interactions between plants and the environment. The resulting stimuli of the environment in which the plant is found can reorient metabolic pathways and influence the synthesis of distinct compounds (Rockenbach et al. 2018).

There are an estimated 3000 known essential oils, among which about 300 are already used in the pharmaceutical, food, and cosmetics industries. Studies have shown these compounds to possess diverse biological effects, such as wound healing, anticarcinogenic, antioxidative, and antimicrobial activity (Kaloustian et al. 2008; Rosato et al. 2018; Ryu et al. 2018).

The majority of essential oils has antimicrobial properties, and act as modulators of microbial resistance to antibiotics. The interests of the scientific community and a greater part of society have grown in a complementary way for more natural products that are considered to be biosafe and of low toxicity. Thus, essential oils are a potential alternative to aid microbial control, which is decreasing with the emergence of multiresistant microbial strains (Lopez-Romero et al. 2015; De Rapper et al. 2016; Basto et al. 2017).

MECHANISMS OF ACTION OF ESSENTIAL OILS

The diversity of applications and uses of essential oils has raised questions about their mechanism of action. However, their complex chemical constitution, with different groups of terpenes, terpenoids, and aromatic compounds, makes it difficult to determine a single reaction cascade that explains the mechanism involved in the biological activities of essential oils. Some studies suggest that most essential oils have high antimicrobial activity because they act on the structure of the bacterial cell wall, altering or inhibiting enzymes essential to microbial metabolism. Other authors have reported that essential oils cause changes in the permeability of the bacterial plasma membrane. These activities may lead to the cessation of the vital basic processes of the cell such as electrochemical adjustment, among others, resulting in the loss of chemoosmotic equilibrium, causing cell death (Carson, Mee and Riley 2002; Aijaz et al. 2011; Bona et al. 2012; Hyldgaard, Mygind and Meyer 2012; Diao et al. 2014; Li et al. 2014; Barbosa et al. 2015; Yang et al. 2015; Zhang et al. 2016; Zhang et al. 2017).

The sensitivity of microorganisms to essential oils has not yet been fully elucidated. The mechanisms of action of oils may be correlated mainly with their composition (minor and major components) as well as with seasonal factors, extraction methods, and storage. Thus, several factors may contribute to the heterogeneity of the data obtained in surveys, causing differences in the reported results (Trombetta et al. 2005; Bachir and Benali 2012; De Rapper et al. 2016).

Although its individual chemical compounds may have been determined to have antimicrobial potential, the impact of an essential oil may be largely due to the interactions between the chemical constituents. Thus, antimicrobial activity is due not only to the major compounds present in the essential oil, but also to the existence of other components in lower concentrations that promote synergistic interactions. p-Cymene, for example, has no antibacterial potential if used alone, but in conjunction with carvacrol, it facilitates the transport of the latter through the cytoplasmic membrane into the bacterial cell. Thus, essential oils are excellent alternative antimicrobials of great potential (Silva et al. 2010; Hyldgaard, Mygind and Meyer 2012).

Among the major constituents of essential oils, phenylpropanoids (phenolic compounds) may be highlighted; they are volatile liposoluble compounds that, in parallel with terpenoids, make up the main substances found in essential oils. They are flavorants whose main precursor is p-coumaric acid; examples are eugenol and cinnamic aldehyde present in high concentrations in the essential oils of cinnamon. Among phenolic compounds, p-coumaric acid is identified as the most active and common plant antioxidant with the ability to eliminate free radicals (Planalto 2016).

Phenolic compounds consist of flavonoids and phenolic acids (benzoic and cinnamic acids and their derivatives). Subclasses of phenols exhibit effects on various proteins involved in metabolic processes, as well as their signaling. Phenylpropanoids perform various functions related to natural defenses in plants against both abiotic (light, temperature, and humidity) and biotic (insect–plant or plant–plant) agents, as well as those involved in internal aspects, such as genetic, nutrients, and hormones, that contributing to its synthesis (Alves et al. 2018).

Other major components are terpenoids, classified as monoterpenes, sesquiterpenes, diterpenes, and triterpenes (Morais 2009) based on the size of the molecule. Monoterpenes and sesquiterpenes are the most abundant terpenoids in nature. Monoterpenes, which contain two isoprene units (C10), are volatile compounds of low molecular weight and characteristic odor. The best-known are limonene (present in citrus fruits) and menthol (present in wild mint and peppermint, among others). Sesquiterpenes are formed by three isoprene units (C15), and are divided into classes, such as bisabolenes, zingiberenes, caryophyllenes, cardinenes, and azulenes. Chamazulene, an azulene derivative, is present mainly in chamomile essential oil produced from the flowers, giving it a blue coloration (Dewick 2009). In addition, sesquiterpenes are also responsible for much of the biological activity of essential oils.

These organic compounds are produced by the plant in order to avoid damage by external agents. Thus, terpenes exhibit recognized antimicrobial activity (Lutfi and Roque 2014; De Martino et al. 2015; Felipe and Bicas 2017). In general, these compounds are responsible for the aroma of many natural products, and, therefore, they have wide versatility, especially in perfumery, cosmetics, pharmacology, and fine chemicals industries (Pavela 2015). In the food industry, terpenoids have contributed to improved sensory characteristics of foods (Ravindra and Kulkarni 2015).

ESSENTIAL OILS AND ANTIBIOFILM PROPERTIES

Several microorganisms are recognized for their ability to adhere to biotic and abiotic surfaces by forming aggregates such as biofilms, considered a mechanism of resistance to antibiotics, that has direct implications for persistence and survival in the environment (Gupta and Birdi 2017; Pontes et al. 2019). Biofilms are surface-associated microbial communities enclosed in a self-generated exopolysaccharide matrix (EPS), which is responsible for adhesion to the surface and cohesion of the biofilm. The EPS accounts for over 90% of the biofilm complex whereas only 10% is formed by the microorganism mass (Flemming and Wingender 2010). Made of biologically active molecules, such as proteins, nucleic acids, lipids, and humic substances, the EPS protects microorganisms against desiccation, oxidizing or charged biocides, some antibiotics and metallic cations, ultraviolet radiation, and the host immune system (Kannappan et al. 2019).

Biofilm development includes five steps: (i) initial attachment of microbial cells to the surface; (ii) production of the EPS, resulting in firm adhesion; (iii) development of early biofilm architecture; (iv) maturation of the biofilm; and (v) dispersal of single cells from the mature biofilm, leading to formation of a new biofilm (Figure 1; Zacchino et al. 2017). Each stage of biofilm production involves aspects of bacterial physiology and associated phenotypic responses influenced by environmental conditions. Natural ecosystems are generally low in available nutrients, and biofilm formation is an important adaptation to survival under these conditions (Tolker-Nielsen 2015).

In the first stage of biofilm formation, the interaction between the microorganism cells and the surface to be colonized is weak, and consists of interacting Van de Waals, electrostatic, and hydrophobic forces. During this adhesion phase, the microorganisms are easily removed with shear force (Chmielewski and Frank 2003). Irreversible binding is established as a result of formation of bacterial appendages and short-range substrate forces such as dipoledipole interaction, hydrogen bonding, and hydrophobic and ionic covalent bonding. Removal of irreversibly attached cells is difficult and requires application of strong shear forces (scrubbing or scraping) or chemical breaking of attachment forces through the application of enzymes, detergents, surfactants, sanitizers, or heat (Chmielewski and Frank 2003).

Microorganisms organized in biofilms are more resistant to antibiotics and chemical agents than cells in suspension. This is due to the three-dimensional architecture of the mature biofilm and the EPS acting as a physical barrier against penetration by antimicrobial agents. To be effective, chemical agents need to penetrate and hydrolyze the biofilm matrix, causing contact between the agent and the site of action of the microorganism in the cell, thus promoting its antimicrobial effect (Bridier et al. 2011; Bazargani and Rohloff 2016; Jardak et al. 2017). Another problem is that the majority of antimicrobial agents are active against microbial cells that do not occur as biofilms (De La Fuente-Núñez et al. 2013).

Natural products such as essential oils have been explored for their action against biofilm formation. Although not clearly defined, the mode of action of their antibacterial activity is attributed mainly to the rupture of the bacterial cytoplasmic membrane, increased permeability and loss of cellular constituents, and modification of a variety of enzymes involved in the production of cellular energy and synthesis of structural components (Bink, 2011; Gupta and Birdi 2017). Antibiofilm activity of essential oils has been reported mostly against *Escherichia coli*, *Klebsiella pneumoniae*, *Lysteria monocytogenes*, *Pseudomonas aeruginosa*, *Staphylococcus aureus*, *S. epidermidis*, and *Salmonella enteritidis*, major species causing infections in humans (Bridier et al. 2011; Jardak et al. 2017). Essential oils, for the most part, prevent initial fixation of bacteria to the surface, reducing adhesion capacity, and, consequently, reducing biofilm formation. This antiadherent action is probably due to the alteration of the proteins in the bacterial surface caused by interaction with the essential oil (Nostro et al. 2007; Čabarkapa et al. 2019).

Another antibiofilm mechanism of essential oils is interference in quorum sensing, which implies an alteration in microbial communication leading to a reduction in expression of virulence factors and formation of biofilms (Inamuco et al. 2012; Bouyahya et al. 2019). Quorum sensing is characterized by chemical signaling between microorganisms and the ability to induce diverse alterations, such as regulation of gene expression dependent on cell density, leading to the formation of new phenotypes (Solano, Echeverz and Lasa 2014). The inhibition of quorum sensing by essential oils is attributed to bioactive compounds that specifically target the quorum sensing signaling pathway implicated in biofilm formation (Jaramillo-Colorado et al. 2012; Bouyahya et al. 2017).

However, application of essential oils as the active principal has its greatest challenge in the complex chemical nature of the oils. Essential oils are unstable and volatile compounds that once extracted from their plant matrix and exposed to factors such as light, air, and temperature variation, can undergo several chemical reactions: isomerization, oxidation, polymerization, and dehydrogenation. These reactions lead to a decrease or loss of their properties (Turek and Stintzing 2013; Majeed et al. 2015).

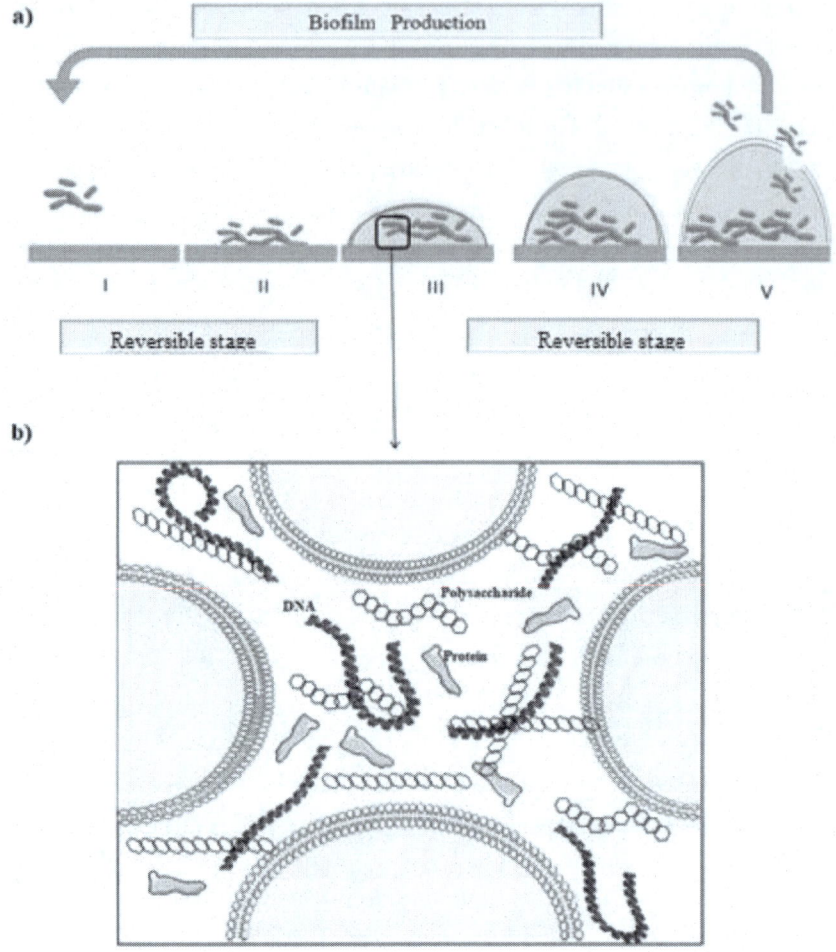

Figure 1. (a) Biofilm development: I. initial attachment of microbial cells to the surface; II. production of the exopolysaccharide matrix (EPS), resulting in firm adherence; III. development of early biofilm architecture; IV. maturation of the biofilm; and V. dispersal of single cells from the mature biofilm, leading to the formation of a new (b) biofilm EPS structure.

ENCAPSULATION AND STABILITY OF ESSENTIAL OILS

To ensure the functionality and shelf life of essential oils, researchers have turned to encapsulation techniques that aim to protect the compounds from oxidation and control their release, thus preserving their stability and prolonging performance. Sagalowicz and Leser (2010) define encapsulation as the process in which the bioactive of interest is enveloped by a coating or aggregated into a matrix forming small particles. This process can be physical, chemical, or physico-chemical, and results in structures such as liposomes, micelles, microparticles, nanoparticles, emulsions, and even encapsulation within yeasts (Rodriguez et al. 2016).

One of the most widely used physical processes for encapsulation in the food industry is called spray drying. This technique is considered fast and low cost and capable of producing spherical microcapsules with homogeneous distribution (Rodriguez et al. 2016). It consists of preparation of an emulsion containing the bioactive of interest, the substance that acts as a "wall" coating the microcapsule, and a solvent. This emulsion will be dried and the solvent evaporated. The reduction in exposure of the bioactive to heat preserves the characteristics of the active principals (Fernandes et al. 2017).

Tomazelli et al. (2018) conducted a study on microencapsulation of essential oil of thyme and its antibacterial activity on *Vibrio alginolyticus* and *V. parahaemolyticus*, pathogenic microorganisms that are responsible for 100% mortality of farm-raised shrimp. The microencapsulation was performed with maltodextrin as wall material, and had an efficiency of encapsulation of 87.16%, obtaining capsules with a mean of 10.37 ± 0.0184 μm. The results showed that antimicrobial activity of the essential oil remained the same in its encapsulated form making this technique feasible for preservation of its stability.

Among the various chemical processes for encapsulation of materials, the *in situ* polymerization technique is considered the most common and easiest to perform. In heterogeneous systems, such as those involving essential oils, *in situ* polymerization can be performed by interfacial polymerization, suspension polymerization or emulsion polymerization. Preparation of emulsions is generally the basis for other techniques (Fernandes et. al. 2018). Emulsions are generally defined as colloidal dispersions of uniform small particles in a particular fluid in which they are not soluble. The particles are formed by a stabilizing emulsifying agent, and have an internal or dispersed phase and an external or dispersing phase (Basto et. al. 2017).

Essential oils can be encapsulated in droplets in an oil-in-water (O/W) emulsion system. However, for better physical stability, these droplets must be at a nanometer scale (20200 nm). Nanoemulsions have a low concentration of surfactant (3–10%) and better intracellular penetration in biological tissues. This results in slow release of the bioactive compounds, reducing the risk of irritability (Bouchemal et al. 2004). Essential oil-based nanoemulsions are widely studied for their antimicrobial potential in the areas of food preservation and production of new drugs (Hilbig et al. 2016; Li et. al. 2017; Chuesiang et. al. 2019).

Gündel et al. (2018) evaluated the action of a nanoemulsion containing the essential oil of *Cymbopogon flexuosus* (lemongrass), produced by high agitation homogenization, on pathogenic microorganisms and their biofilms. The nanoemulsion presented uniformly distributed droplets smaller than 200 nm, and maintained the same physical-chemical parameters for 90 days under refrigeration, proving its stability. Nanoencapsulation of the essential oil also increased its therapeutic efficacy against *Cryptococcus grubii*, *Staphylococcus aureus*, and *Candida albicans* when compared to the free oil.

Among physical-chemical processes, the coacervation technique, also known as the phase separation method, has been highlighted; it is a simple method that allows the production of microencapsulated particles. This technique is based on the phenomenon in which a liquid phase separates from a macromolecular solution, as the result of a reduction in solubility by chemical or physical means, creating a new phase that is rich in colloids that appear in the form of liquid droplets, the coacervate. These droplets agglutinate and form the walls of the capsules. Subsequently, the walls are hardened and the capsules are separated (Azeredo 2008).

Peng et al. (2014) tested the antimicrobial action of white mustard (*Sinapis alba*) seed essential oil, and verified its chemical composition and stability when microencapsulated by a complex coacervation process with genipin, a natural crosslinking agent used to form the walls of microcapsules. The study showed that the essential oil had antimicrobial action against *Staphylococcus aureus, S. epidermidis, Micrococcus luteus, Escherichia coli, Bacillus subtilis, Shigella sonnei, Salmonella lignieres, Pseudomonas aeruginosa*, and *P. fluorescens*. In addition, the microcapsules obtained preserved the stability of the essential oil at a relative humidity of 54%, with an oil retention rate of 85.74%.

The preservation of the structure of sensitive molecular compounds such as essential oils is one of the major contributions of the advancement of encapsulation technology. The exponential increase in the applicability of essential oils is of great importance for science as a whole.

CONCLUSION

Based on the available literature, it is possible confirm that most essential oils present antimicrobial activity against planktonic cells as well as against those organized in biofilm. The main advantages of synthetic biocides are low action, low toxicity or irritability, and modulation of bacterial resistance, reducing the appearance of multiresistant strains. Because they are products of plant origin, essential oils are influenced by environmental factors (climate, temperature, humidity, rain, sunlight, etc.) and the techniques used for their extraction. These variations must be considered when conducting research or in industrial applications, since they may result in changes to the yield of the oils and to the major components, and, consequently, to the bioactivity of the product. In addition, it is also necessary to study the stability of the oils during their time of commercialization, since several factors can promote changes in their components leading to inactivation of the desired properties. To improve the stability of essential oils and increase their shelf life, encapsulation techniques have been developed. Heterogeneity in the quality of the final product is one of the factors that limit its industrial application, and studies must be carried out to improve extraction standards for the oils, which will allow improved acceptability for different applications of these compounds.

REFERENCES

Ahmad, Aijaz., et al. (2011). "Antifungal activity of *Coriaria nepalensis* essential oil by disrupting ergosterol biosynthesis and membrane integrity against Candida." *Yeast.*, *28*, 611-617. Doi: 10.1002/ yea.18 90.

Ali, Babar., et al. (2015) "Essential oils used in aromatherapy: A systemic review." *Asian Pacific Journal of Tropical Biomedicine.*, *5*, 601-611. Doi: 10.1016/j.apjtb.2015.05.007.

Alves, Vanessa., et al. (2018). "Sensory acceptability and physico-chemical haracteristics of dehydrated strawberries with different treatments." *Demetra: Food, Nutrition & Health.*, *13*, 745-764. Doi: 10.12957/demetra.2018.31920.

Astani, Akram., et al. (2010). "Comparative study on the antiviral activity of selected monoterpenes derived from essential oils." *Phytotherapy Research: of Natural Product Derivatives.*, *24*, 673-679. Doi: 10.1002/ ptr.2955.

Auddy, B., et al. (2003). "Screening of antioxidant activity of three Indian medicinal plants, traditionally used for the management of neurodegenerative diseases." *Journal of Ethnopharmacology.*, *84*, 131-138. Doi: 10.1016/S0378-8741(02)00322-7.

Azeredo, Henriete. M. C. (2008). "Encapsulation: Application to Food Technology." *Alimentos e Nutrição Araraquara*, *16*, 89-97.

Bachir, Raho G. & Benali, M. (2012). "Antibacterial activity of the essential oils from the leaves of Eucalyptus globulus against *Escherichia coli* and *Staphylococcus aureus*." *Asian Pacific Journal of Tropical Biomedicine*, *2*, 739-742. Doi: 10.1016/S2221-1691(12)60 220-2.

Barbosa, Lidiane Nunes., et al. (2015). "*In vitro* antibacterial and chemical properties of essential oils including native plants from Brazil against pathogenic and resistant bacteria." *Journal of Oleo Science.*, *64*, 289-298. Doi: 10.5650/jos.ess14209.

Barros Luz, Janaina., et al. (2015). "The use of açaí as a potential antioxidant for bovine semen cryopreservation." *Medicina Veterinária UFRPE, 12,* 143-153: Doi: 10.26605/medvet-v12n2-2366.

Basto, S. R. L., et al. (2017). "Emulsion and microemulsion: new controlled release systems." *Medicina Veterinária UFRPE, 10,* 25-33. 10.26605/medvet-v2n2-3466.

Bazargani, Mitra Mohammadi Rohloff Jens. (2016). "Antibiofilm activity of essential oils and plant extracts against *Staphylococcus aureus* and *Escherichia coli* biofilms." *Food Control, 156,* 164-61. Doi 10.1016/j.foodcont.2015.09.036.

Benjilali, B., et al. (1986). "Method of study of the antiseptic properties of essential oils by direct contact in agar medium." *Plantes Médicinales et Phytothérapie, 20,* 155-167.

Bink, A., et al. (2011). "Anti-Biofilm Strategies: How to Eradicate Candida Biofilms?" *The Open Mycology Journal., 5,* 29-38. Doi: 10.2 174/1874437001105010029.

Bona, Tânia. D. M. M., et al. (2012). "Oregano, rosemary, cinnamon and pepper extract essential oil to control Salmonella, Eimeria and Clostridium in broilers." *Pesquisa Veterinária Brasileira., 35,* 411-418.Doi: 10.1590/S0100-736X2012000500009.

Bouchemal, K., et al. (2004). "Nano-emulsion formulation using spontaneous emulsification: solvent, oil and surfactant optimisation." *International journal of pharmaceutics., 280,* 241-251. Doi: 10.1016/j.ijpharm.2004.05.016.

Bouyahya, Abdelhakim., et al. (2017). "Medicinal plant products targeting quorum sensing for combating bacterial infections". *Asian Pacific Journal of Tropical Medicine., 10,* 729-743. Doi: 10.1016/j.apjtm.2 017.07.021.

Bouyahya, Abdelhakim., et al. (2019). "Essential oils of Origanum compactum induce membrane permeability, disturb cell membrane, and suppress quorum-sensing phenotype in bacteria." *Journal of Pharmaceutical Analysis, 1,* 11. Doi: 10.1016/j.jpha.2019.03.001.

Bridier, A., et al. (2011). "Biofouling: The Journal of Bioadhesion and Biofilm Resistance of bacterial biofilms to disinfectants: a review." *Biofouling.*, *27*, 37-41. Doi: 10.1080/08927014.2011.626899.

Čabarkapa, Ivana., et al. (2019). "Anti-biofilm activities of essential oils rich in carvacrol and thymol against *Salmonella Enteritidis*." *Biofouling.* 1–15. Doi: 10.1080/08927014.2019.1610169.

Carson, Christine F., et al. (2002). "Mechanism of action of *Melaleuca alternifolia* (tea tree) oil on *Staphylococcus aureus* determined by time-kill, lysis, leakage, and salt tolerance assays and electron microscopy." *Antimicrobial Agents and Chemotherapy.*, *6*, 1914-1920. Doi: 10.1128/AAC.46.6.1914-1920.2002.

Carvalho, Filho., José Luiz, S., et al. (2006). "Influence of the harvesting time, temperature and drying period on basil (*Ocimum basilicum* L.) essential oil." *Revista Brasileira de Farmacognosia.*, *30*, 16-24. Doi: 10.1590/s0102-695x2006000100007.

Chmielewski, R. A. N. & Frank, J. F. (2003). "Biofilm Formation and Control in Food Processing Facilities." *Comprehensive Reviews in Food Science and Food Safety.* Doi: 10.1111/j.1541-4337.2003.tb00012.x.

Chouhan, S., et al. (2017). "Antimicrobial activity of some essential oils—present status and future perspectives." *Medicines.*, *4*, 58. Doi: 10.3390/medicines4030058.

Chuesiang, Piyanan., et al. (2019). "Antimicrobial activity and chemical stability of cinnamon oil in oil-in-water nanoemulsions fabricated using the phase inversion temperature method." *LWT.*, *514*, 208-216. Doi: 10.1016/j.jcis.2017.11.084.

Costa, D. Carvalho., et al. (2015). "Advances in phenolic compounds analysis of aromatic plants and their potential applications." *Trends in Food Science & Technology.*, *45*, 366-354. Doi: 10.1016/ j.tifs.2015.06.009.

Costa, J. G. M., et al. (2005). "Chemical-biological study of the essential oils of Hyptis martiusii, Lippia sidoides and *Syzigium aromaticum* against the larvae of *Aedes aegypti*." Revista Brasileira de

Farmacognosia, *15*(4), 304-309. Doi: 10.1590/S0102-695X2005000 400008.

Costa, Patrícia Silva., et al. (2017). "Antimicrobial activity and therapeutic potential of the genus *Lippia sensu* lato (Verbenaceae)." *Hoehnea*. 44:158-171.Doi: 10.1590/2236-8906-68/2016.

Da Silva Gündel, Samanta., et al. (2018). "Nanoemulsions containing Cymbopogon flexuosus essential oil: Development, characterization, stability study and evaluation of antimicrobial and antibiofilm activities." *Microbial pathogenesis.*, *118*, 268-276. Doi: 10.1016/j.micpath.2018.03.043.

De Billerbeck, V. G. (2007). "Essential oils and bacteria resistant to antibiotics." *Phytothérapie.*, *249*, 253-5. Doi: 10.1007/s10298-007-0265-z.

De la Fuente-Núñez, César., et al. (2013). "Bacterial biofilm development as a multicellular adaptation: Antibiotic resistance and new therapeutic strategies." *Current Opinion in Microbiology.*, *16*, 580-589. Doi: 10.1016/j.mib.2013.06.013.

De Rapper, Stephanie., et al. (2016). "The *in vitro* antimicrobial effects of Lavandula angustifolia essential oil in combination with conventional antimicrobial agents." *Evidence-Based Complementary and Alternative Medicine.*, 1-9. Doi: 10.1155/2016/2752739.

Diao, Wen Rui., et al. (2014). "Chemical composition, antibacterial activity and mechanism of action of essential oil from seeds of fennel (*Foeniculum vulgare* Mill.)." *Food Control.*, *35*, 119-116. Doi: 10.1016/j.foodcont.2013.06.056.

Dudareva, Natalia. & Pichersky, Eran. (2008). "Metabolic engineering of plant volatiles." *Current Opinion in Biotechnology.*, *19*(2), 181-189. Doi: 10.1016/j.copbio.2008.02.011.

Fernandes, Iara. J. et al. (2018). "Production and evaluation of alginate microcapsules containing orange peel essential oil." *Eclética Química Journal.*, *39*, 164-174. Doi: 10.26850/1678-4618eqj.v39.1. 2014.p164-174.

Fernandes, Regiane., et al. (2017). "Behavior of microencapsulated rosemary essential oil by spray drying at different relative humidity." *Revista Ciência Agrícola, 14*, 73-82. Doi: 10.28998/rca.v14i1.2469.

Ferreira, S. D., et al. (2016). "Effect of nitrogen fertilization and seasonality on productivity of Ocimum basilicum L." *Revista Brasileira de Plantas Medicinais, 18*(1), 67-73. Doi: 10.1590/1983-084X/15_035.

Fisher, Katie. & Phillips, Carol. (2018). "Potential antimicrobial uses of essential oils in food: is citrus the answer?" *Trends in Food Science & Technology., 156*, 164-19. Doi: 10.1016/j.tifs.2007.11.006.

Flemming, Hans-Curt. & Wingender, Jost. (2010). "The biofilm matrix. Nature." *Reviews Microbiology., 623*, 633-8. Doi: 10.1038/nrmicro 2415.

Gupta, Pooja D. & Birdi, Tannaz J. (2017). "Development of botanicals to combat antibiotic resistance." *Journal of Ayurveda and Integrative Medicine., 8*, 266-275. Doi: 10.1016/j.jaim.2017.05.004.

Hajlaoui, Hafedh., et al. (2010). "Chemical composition and biological activities of Tunisian *Cuminum cyminum* L. essential oil: A high effectiveness against Vibrio spp. strains." *Food and Chemical Toxicology., 48*(8-9), 2186-2192. Doi: 10.1016/j.fct.2010.05.044.

Hilbig, Jonas., et al. (2016). "Physical and antimicrobial properties of cinnamon bark oil co-nanoemulsified by lauric arginate and Tween 80." *International journal of food microbiology., 233*, 52-59, Doi: 10.1016/j.ijfoodmicro.2016.06.016.

Hyldgaard, Morten., et al. (2012). "Essential oils in food preservation: mode of action, synergies, and interactions with food matrix components." *Frontiers in Microbiology.* Doi: 10.3389/fmicb.2012.00012.

Inamuco, Jeanine., et al. (2012). "Sub-lethal levels of carvacrol reduce *Salmonella Typhimurium* motility and invasion of porcine epithelial cells." *Veterinary Microbiology., 157*, 200-207.Doi:10.1016/j.vetmic.2011.12.021.

Jaramillo-Colorado, Beatriz., et al. (2012). "Anti-quorum sensing activity of essential oils from Colombian plants." *Natural Product Research.*, *26*, 1075-1086. Doi: 10.1080/14786419.2011. 557376.

Jardak, Marwa., et al. (2017). "Chemical composition, anti-biofilm activity and potential cytotoxic effect on cancer cells of *Rosmarinus officinalis* L. essential oil from Tunisia." *Lipids in Health and Disease.*, *16*, 1-10. Doi: 10.1186/s12944-017-0580-9.

Jentzsch, Paul., et al. (2015). "Handheld Raman spectroscopy for the distinction of essential oils used in the cosmetics industry." *Cosmetics.*, *2*(2), 162-176. Doi: 10.3390/cosmetics2020162.

Kaloustian, J., et al. (2008). "Study of six essential oils: chemical composition and antibacterial activity." *Phytothérapie*, *6*, 160-164. Doi: 10.1007/s10298-008-0307-1.

Kannappan, Arunachalam., et al. (2019). "*In vitro* and *in vivo* biofilm inhibitory efficacy of geraniol-cefotaxime combination against Staphylococcus spp." *Food and Chemical Toxicology.*, *125*, 322:332. Dio: 10.1016/j.fct.2019.01.008.

Krishna, A., et al. (2000). "Aromatherapy-an alternative health care through essential oils." *Journal of Medicinal and Aromatic Plant Sciences.*, *22*, 798-804.

Lambert, R. J. W., et al. (2001). "A study of the minimum inhibitory concentration and mode of action of oregano essential oil, thymol and carvacrol." *Journal of Applied Microbiology.*, *9*, 453-462. Doi: 10.1046/j.1365-2672.2001. 01428.x.

Li, Jianming., et al. (2017). "Thymol nanoemulsions formed via spontaneous emulsification: Physical and antimicrobial properties." *Food chemistry.*, *232*, 191-:197. Doi: 10.1016/j.foodchem.2017.03. 147.

Li, Wen Ru., et al. (2014). "Antibacterial activity and kinetics of *Litsea cubeba* oil on *Escherichia coli*." *Plos One.*, *9*(11), e110983. Doi: 10.1 371/journal.pone.0110983.

Li, Zhi Jian., et al. (2013). "Chemical composition and antimicrobial activity of the essential oil from the edible aromatic plant *Aristolochia delavayi.*" *Chemistry & Biodiversity.*, *10*(11), 2032-2041. Doi: 10.1002/cbdv.201300066.

Lima, Igara de Oliveira., et al. (2006). "Antifungal activity of essential oils on Candida species." *Brazilian Journal of Pharmacognosy.*, *16*(2), 197-201. Doi: 10.1590/s0102-695x2006000200011.

Lopez-Romero, Julio Cesar., et al. (2015). "Antibacterial effects and mode of action of selected essential oils components against *Escherichia coli* and *Staphylococcus aureus.*" *Evidence-Based Complementary and Alternative Medicine.*, 1-9. Doi: 10.1155/2015/795435.

Lupe, F. A. (2007). Study of the chemical composition of aromatic plants essential oils from the Amazon. Masters dissertation. State University Campinas.

Majeed, Hamid., et al. (2015). "Essential oil encapsulations: uses, procedures, and trends." *RSC Advances.*, *5*, 58449-58463. Doi: 10.1039/c5ra06556a.

Medeiros, Rosane Tamara da Silva., et al. (2011). "Evaluation of antifungal activity of *Pittosporum undulatum* L. essential oil against *Aspergillus flavus* and aflatoxin production." *Ciência e Agrotecnologia*, *35*(1), 71-76. Doi: 10.1590/s1413-70542011000100008.

Nostro, Antonia., et al. (2007). "Effects of oregano, carvacrol and thymol on Staphylococcus aureus and *Staphylococcus epidermidis* biofilms." *Journal of Medical Microbiology.*, *56*, 519-523. Doi: 10.1099/ jmm.0.46804-0.

Pavela, Roman. (2015). "Essential oils for the development of eco-friendly mosquito larvicides: a review." *Industrial Crops and Products.*, *76*, 174-187. Doi: 10.1016/j.indcrop.2015.06.050.

Peng, Chao., et al. (2014). "Chemical composition, antimicrobial property and microencapsulation of Mustard (*Sinapis alba*) seed essential oil by complex coacervation." *Food chemistry.*, *165*, 560-568. Doi: 10.1016/j.foodchem.2014.05.126.

Pontes, Eveline Kelle Ursulino., et al. (2019). "Antibiofilm activity of the essential oil of citronella (*Cymbopogon nardus*) and its major component, geraniol, on the bacterial biofilms of Staphylococcus aureus." *Food Science and Biotechnology.*, *633*, 639. Doi: 10.1007/s10068-018-0502-2.

Ribeiro, Soraya Marques., et al. (2018). "Influence of seasonality and circadian cycle on yield and chemical composition of essential oils of Croton spp. from Caatinga." *Iheringia. Série Botânica*, *73*(1), 31-38. Doi: 10.21826/2446-8231201873104.

Rockenbach, Ana Paula., et al. (2018). "Interference between weeds and crop: changes in secondary metabolism." *Revista Brasileira de Herbicidas*, *17*(1), 59-70. Doi: 10.7824/rbh. v17i1.527.

Rodríguez, Julia., et al. (2016). "Current encapsulation strategies for bioactive oils: From alimentary to pharmaceutical perspectives." *Food Research International.*, *83*, 41-59. Doi: 10.1016/j.foodres.2016. 01.0 32.

Rosato, Antonio., et al. (2018). "Elucidation of the synergistic action of Mentha Piperita essential oil with common antimicrobials." *PloS one.*, *13*(8), e0200902 Doi: 10.1371/journal.pone.0200902.

Ryu, Victor., et al. (2018). "Effect of ripening inhibitor type on formation, stability, and antimicrobial activity of thyme oil nanoemulsion." *Food Chemistry.*, *245*, 104-111. Doi: 10.1016/j.foodchem.2017.10.084.

Safaei-Ghomi, Javad. & Ahd, AtefehAbbasi. (2010). "Antimicrobial and antifungal properties of the essential oil and methanol extracts of Eucalyptus largiflorens and Eucalyptus intertexta." *Pharmacognosy Magazine.*, *6*, 172. Doi: 10.4103/0973-1296.66930.

Sagalowicz, Laurent. & Leser, Martin E. (2010). "Delivery systems for liquid food products." *Current Opinion in Colloid & Interface Science.*, *15*, 61-72. Doi: 10.1016/j.cocis.2009.12.003.

Salvia-Trujillo, Laura., et al. (2015). "Physicochemical characterization and antimicrobial activity of food-grade emulsions and nanoemulsions incorporating essential oils." *Food Hydrocolloids.*, *43*, 547-556. Doi: 10.1016/j.foodhyd.2014.07.012.

Silva, Janine Passos Lima., et al. (2010). "Oregano essential oil: interference of chemical composition on activity against *Salmonella* Enteritidis." *Ciência e Tecnologia de Alimentos.* Doi: 10.1590/s0101-20612010000500021.

Solano, C., et al. (2015). "Biofilm dispersion and quorum sensing." *Current Opinion in Microbiology.*, *96*, 104-18. Doi: 10.1016/j.mib.2014.02.008.

Tolker-Nielsen, Tim. (2015). "Biofilm Development." *Microbiology Spectrum.*, *32*, 3-252. Doi: 10.1128/microbiolspec. MB-0001-2014.

Tomazelli Júnior, O., et al. (2018). "Microencapsulation of essential thyme oil by spray drying and its antimicrobial evaluation against *Vibrio alginolyticus* and *Vibrio parahaemolyticus*." *Brazilian Journal of Biology.*, *78*, 311-317.Doi: 10.1590/1519-6984.08716.

Trombetta, Domenico., et al. (2005). "Mechanisms of antibacterial action of three monoterpenes." *Antimicrobial Agents and Chemotherapy.*, *47*, 2474-2478. Doi: 10.1128/AAC.49.6.2474-2478.2005.

Turek, Claudia. & Stintzing, Florian C. (2013). "Stability of essential oils: a review." *Comprehensive Reviews in Food Science and Food Safety.*, *12*, 40-53. Doi: 10.1111/1541-4337.12006.

Yang, Xiao Nan., et al. (2015). "Chemical composition, mechanism of antibacterial action and antioxidant activity of leaf essential oil of *Forsythia koreana* deciduous shrub." *Asian Pacific Journal of Tropical Medicine.*, *8*, 694-700 Doi: 10.1016/j.apjtm.2015.07.031.

Zacchino, Susana A., et al. (2017). "Hybrid combinations containing natural products and antimicrobial drugs that interfere with bacterial and fungal biofilms." *Phytomedicine.*, *37*, 14-26. Doi: 10.1016/j.phymed.2017.10.021.

Zhang, Jing., et al. (2017). "Antibacterial activity and mechanism of action of black pepper essential oil on meat-borne *Escherichia coli*." *Frontiers in Microbiology.*, *7*, 2094. Doi: 10.3389/fmicb.2016.02094.

Zhang, Yunbin., et al. (2016). "Antibacterial activity and mechanism of cinnamon essential oil against *Escherichia coli* and *Staphylococcus aureus*." *Food Control.*, *59*, 282-289. Doi: 10.1016/ j.foodcont.2015.05.032.

In: Antimicrobial Potential of Essential Oils ISBN: 978-1-53616-945-4
Editors: B. Oliveira de Veras et al. © 2020 Nova Science Publishers, Inc.

Chapter 2

ANTIBACTERIAL ACTIVITY OF ESSENTIAL OILS IN AROMATHERAPY PROTOCOLS

Yago Queiroz dos Santos[1,2*],
Gabriella Silva Campos Carelli[2],
Anderson Felipe Jácome de França[2],
Clecia de Carvalho Marques[3],
Marcia Vanusa da Silva[3],
Maria Betânia Melo de Oliveira[3],
Ana Catarina de Souza Lopes[4]
and Bruno Oliveira de Veras[4]

[1]Institute of Tropical Medicine Natal, Rio Grande do Norte, Brazil
[2]Federal University of Rio Grande do Norte,
Natal, Rio Grande do Norte, Brazil
[3]Federal University of Pernambuco, Recife-PE, Brazil
[4]Federal University of Pernambuco, Recife, Pernambuco, Brazil

---

* Corresponding Author's Email: yagoqs@hotmail.com.

Abstract

Aromatherapy is the therapy that employs plantsvolatile aromatic elements, that is, their essential oils. The practice that helps patients to sleep and rest as well as helps on alertness, creativity, among others. The therapeutic massage integrates a set of practices with maneuvers whose goal is to promote health and balance with the body, promoting psychological effects on the skin, visceral pain. This pharmacological components of essential oils are volatile constituents at temperature environment, most of which originate from secondary metabolism produced and stored in their own secretory structures formed in the leaves, flowers, branches, stems or roots of various species usually secreted by glandular trichomes, which have various forms, structures and functions distributed mainly on the surface of the leaves. The identification of essential oil constituents is important for the understanding and prediction of their physiological effects where main studied activities are antimicrobial activity, namely antibacterial, antifungal and antiviral, anxiolytic, antioxidant, anticarcinogenic and antinociceptive consisting a helpful and potential pharmacological sources in order to develop new clinical aromatherapy protocols.

Introduction

Aromatherapy consists of a branch of Phytotherapy that is based on the premise that essential oils acts like therapeutic agents in promoting and maintaining the well-being of the individual, who goes through the treatment and prevention of diseases, as well as in the treatment of altered emotional states (Coelho 2009; Naha 2014). This therapy is seen as a complementary medicine, used mostly in primary health care being compatible with classical therapy (Cunha and Roque 2013). According to Jade Shutes (Naha 2014) aromatherapy is a natural and non-invasive practice, designed to not only act on the symptom or disease, but also to maintain the natural balance of the organism as a whole, by the correct use of the essential oils. Such a definition gives the holistic view of this therapy, that is, it addresses the whole body including the physical and mental part of the individual (Naha 2014). In India, aromatherapy has been practiced for 6000 years and is still widely used nowadays through

Ayurvedic Medicine, which includes massage with oils (Cunha and Roque 2013; AIA 2014).

It is estimated that in China, its employment may have arisen even earlier than in India. In ancient Egypt, essential oils were used in religious practices associated with the treatment of diseases as well as onmummification of corpses, as an attempt to keep intact the body so its soul could return for the afterlife. They were still used to perfume temples and dwellings and as offerings to the gods. These practices spread to neighboring civilizations, such as Greece and Italy (Cunha and Roque 2013; AIA 2014) that have been adopting applications.

Hippocrates, a Greek known as the father of medicine, refers in his manuscripts to both aromatic substances and their use in massages, praising the physician's role in these practices. According to him "the key to good health lies to take an aromatic bath and a massage with essences per day "(Cunha and Roque 2013; EIA 2014). One of the great steps for the evolution of the use of essential oils was the development of the technique of extraction by distillation by the Arab countries in the century X (AIA 2014). Subsequently, essential oils arrived in Europe by hand of medieval knights in the twelfth century (Cunha and Roque 2013). The seventeenth and eighteenth centuries were marked in the history of aromatherapy as the best age aver to this kind of therapy. To the extent that certain essential oils, such as wormwood, rosemary, nutmeg, garlic, camphor, were used as antiseptics against the pests that haunted this period of history. Thus, in the eighteenth-century essential oils became part of the therapeutic options along with other standard medications. However, as consequence of the appearance of chemically synthesized drugs in the second half of this century. After such discovery, it was preferable to isolate the active principles of the plants and producing analogous chemicals at the expenses of the usage of such natural compounds (Cunha and Roque 2013).

The practice of aromatherapy is based on the responsibility and awareness of possibility of occurrence of undesirable effects associated with the use of essential oils. Several aroma therapists and laymen consider that the essential oils obtained from aromatic plants (Figure 1) are completely safe (Cunha et al. 2012). Its toxicity is often than the one shown in the plant containing them, in so far as they are more concentrate and have the ability to cross biological membranes because of their high lip solubility.

Due to their complex composition, the essential oils for in addition to having several health benefits, may also represent certain scratches. However, with proper use, the potential risks reduced (Cunha et al. 2012). The most frequent undesired reactions are reactions due to direct contact of the essential oils with the skin - dermal reactions - among which are the irritation, sensitization, phototoxicity and photosensitivity (Naha 2014).

Before starting therapy, it is of great importance to evaluate the person who is going aromatherapy, that is, to check if it has some kind of sensitivity or other pathologies. For example, aromatherapy in children can be done, but in a very responsible because it can create risks because of the greater susceptibility to chemical compounds (Figure 2). In the literature it is often found that treatment of cramps in infants and children can deliver 5 to 10 drops of chamomile essential oil with milk. However, there is no reference to either the botanical to the exact amount of oil, and can easily approach undelivered doses (Lis-Balchin 2010).

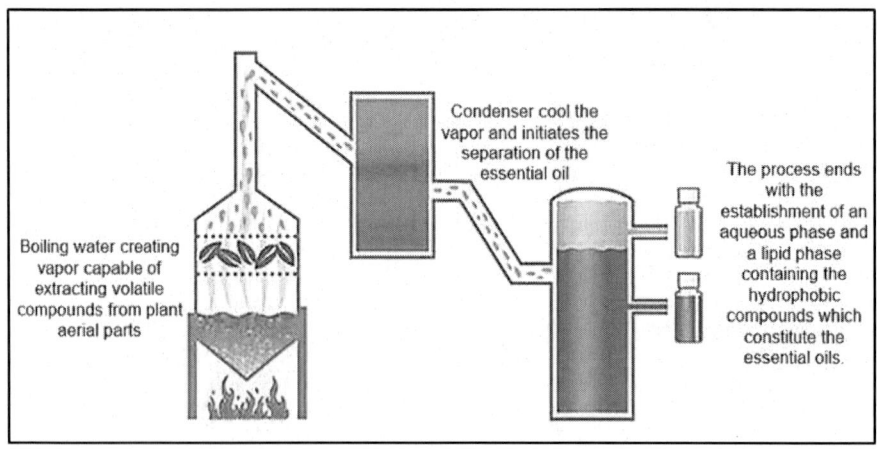

Figure 1. Basic steps of essential oils production.

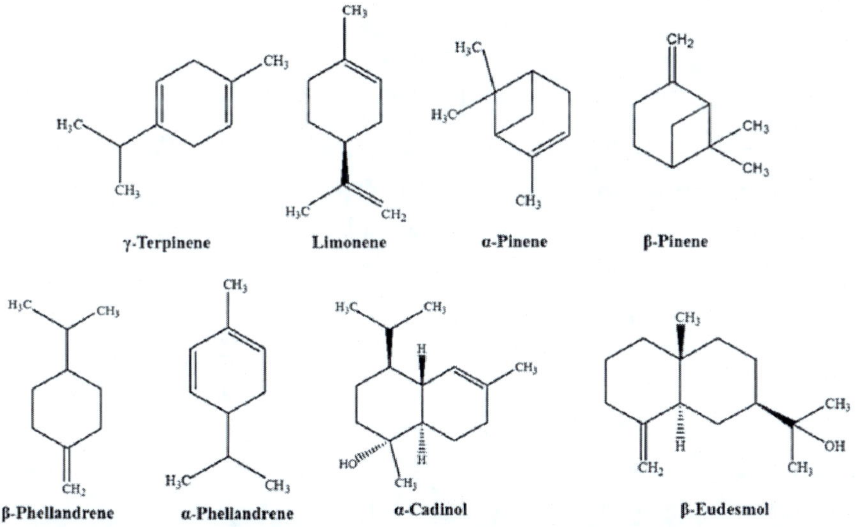

Figure 2. Major molecular constituents of essential oils.

There are several essential oils that are contraindicated for children under the age of six due to the possibility of toxicity reactions such as: *Artemisia arborescens*, *Lavandula latifolia*, *Rosmarinus officinalis* chemotypes (camphor and verbenone, because they have ketones in their composition); *Pimpinela anisum* and *Illicium verum* because they contain anethole, *Cinnamomum ceuylanicum*, *Satureja hortensis* and all species of *Thymus* genus because they may reveal hepatotoxicity due to phenols, and also *Acorus albidum*, *Myrtisticum fragans*, *Petroselium sativum* and *Sassafras albidum* (Cunha et al. 2012).

Aromatherapy in pregnancy is the subject of great controversy. The number of women seeking their own natural products as an alternative medicinal product with no adverse effects. However, oil therapy essential can be extremely unsafe for both the pregnant and the baby, being important to seek information from competent professionals, avoiding self-medication. Essential oils such as aniseed, star anise, calamella, citronella, cumin, tarragon, fennel, hyssop, lábdano, levísco, myrrh, marjoram, parsley, thyme, jasmine, juniper, peppermint, cloves, cedar, sage, rosemary and coriander should be avoided at this stage, with some having abortive properties. Also, lightening, fennel and incense should be avoided because they are emmenagogues hebrs-derived products (Cunha et al. 2012; Sibbritt et al. 2014). People with certain pathologies should also avoid certain essential oils. Hypertensive and diabetic individuals should not use essential oils of rosemary, hyssop and individuals who have epilepsy should avoid the essential oil of fennel and eucalyptus (Cunha et al. 2012).

In general, aromatherapy techniques are divided into two types of application: by inhalation route ortopical application. The first one uses the properties volatile and low molecular weight essential oils so that odors can reach up to receptors of mucosal cells, reach the limbic system and the hypothalamus that are responsible for sensory and motor activities. The essential oils will activate the production of neurotransmitters, such as serotonin, acetylcholine, noradrenaline, endorphins and others that make the communication with all the systems of the organism.

The Inhalation can be done using diffusers, sprays, tissues moistened in oils candles, among other objects. With the power to act in the body so global, acting at the level of the nervous system, there are several types of oils classification. Airey (2004) proposes a classification of the aromas according to the effects by their inhalation (Cunha et al. 2012).

Some researchers present some examples of this classification. The topical route is also widely used. This allows a systemic action due, once again, to the reduced molecular weight and also to its high liposolubility, which allows its solubility in the hydrolipidic film of the skin, as well as in the products secreted by the sweat and/or sebaceous glands, eventually penetrating the different layers of the skin to the hypodermis, where they are absorbed into the bloodstream.

The well-known approach to topical aromatherapy application is through massages on the whole body, which promote a better absorption of increase blood flow to the skin surface and lymphatic drainage. It's also frequent use of cold or hot compresses. Cold compresses are used in cases of bruising, swelling, headaches and fever while hot compresses are recommended for the reduction of cramping abscesses, rheumatic muscles. In both massages and compresses, the essential oil is diluted in a vegetable oil (Cunha and Roque 2013; AIA 2014). In general, the use of essential oils in therapy requires their dilution, since they are very active substances which in such quantities may cause therefore it is common to dilute in carrier oils.

Another common practice in aromatherapy is to immerse the relaxation and reduction of pain, adding essential oils that will enhance the intended effect. Such baths are called aromatic baths. Before they added in the bath water, the oils should be diluted in glycerin or propylene glycol and the water should be warm (approximately 30 ° C). However, if what you want is a bath, the temperature can be 37 ° C, since this temperature reduces muscle tension and dilate blood vessels. It is important that those baths are contraindicated in people with cardiac complications and with varicose veins. Within the category of the aromatic baths can be introduced where there is only the immersion of the feet to the gastrocnemius, known in the slang as "Belly of the leg" at a temperature of around 40 °C. This technique allows dilation of the blood vessels of the lower limbs and are recommended in situations of insomnia, congestion, headaches, minorities, flu and colds (Cunha and Roque 2013).

It's also possible to include in this category facial sprays, where one to three drops of essential oil with antiseptic and soothing properties for the skin without irritants, for example, chamomile, lavender, geranium and also incorporate these oils into clay masks (AIA 2014). The ingestion of essential oils is also possible and can be done by deposition of essential oils drops on sugar cubes or incorporated into honey. At the pharmacy level the essential oils can be applied directly to gelatinous enteric capsules. Due to the high concentration of oils and the risks that may be associated with its intake it is recommended not to exceed the amount of a drop per ten kilograms of body weight (Cunha et al. 2012).

Essential oils are capable of promoting in stressed patients some moments of reflection and relaxation, so that we can remember that we are people with feelings and emotions, as well as our immense intellectual and working potential (Maluf 2008). Essential oils are substances of high volatile power and variable fragrances. These oils are intended to provide relief, healing, stimulation and relaxation. The choice of the type of extraction depends on the nature of the plant material in the natural state, which can be done through several methods, the most used being the distillation method (Hoare 2010).

Lavender oil is considered effective in treating over seventy health problems by having antidepressant, anti-inflammatory, antibacterial, relaxing, sedative, decongestant, toning, antiviral, and soothing properties. Lavenders have pointed leaves and bluish-purple flowers, are non-toxic, and can then be used on the skin without diluting the essential oil, but carefully applying the area to be treated. It is not indicated for pregnant women during the first trimester of pregnancy (Hoare 2010). According to Lee et al. (2006), lavender essential oil has been shown to be effective in insomnia, improving sleep quality and acting on depression in women. The aromas are our closest contact with nature and have the power to predispose us to sleep, rest, alertness, creativity, irritability and creation, among others, since smell is the oldest and perhaps the most unknown of the senses developed by man (Corazza 2002). Receptor cells for the sense of smell are olfactory cells, originating from the central nervous system itself. They are olfactory hair or eyelashes, located at the mucosal end of the olfactory cell, which react to odors present in the air and stimulate olfactory cells (Guyton 1993).

Since olfactory receptor cells are also the primary afferent neurons, their replacement from the basal cells is continuous (neurogenesis). It should be noted that olfactory receptor cells are the only neurons in the adult human being with constant renewal. At the end of the process, the nerve impulses travel through nerve fibers, called axons, until reaching its final destination, the brain, or rather the part of the brain responsible for olfaction, the limbic system. From this moment, the messages will be codified and will produce reactions of physiological and psychological order (Serrano 1985; Guyton 1993; Constanzo 1999 and Corazza 2002).

It is in the limbic system that the cells that process the information coming from the nerve terminals connected to the olfactory bulb (Grace 1999) are located. The first evidence of limbic system involvement in emotional processes arose in 1933, when Herrick stated that the limbic system could have an influence on the organism's affective mechanisms. This idea was further developed by Papez, Mclean and Arnold in 1937,

1949 and 1945, respectively (Corazza 2002). According to Ballone (2009), the agitation of modernity points out that the modern age can be characterized as the Age of Anxiety.

The usage of techniques in order to alleviate the stress of modern life has been increasingly common. For Vera and Vila (2002), these techniques constitute a set of intervention procedures that may be useful for applied psychology in general. Progressive relaxation aims to lead the subject to a deep state of muscular relaxation, in order to provide the psychological and physical well-being in relation to the relation of the emotional state with the corporal.

Benson (1977) found that during relaxation there is a decrease in the heart rate, breathing rate, blood pressure and the amount of oxygen consumed by the body, and therefore, there are hormonal changes and changes in brain electrical waves, with predominance of slow, alpha and theta.

PHARMACOLOGICAL PROPERTIES OF ESSENTIAL OILS

Essential Oils (EOs) are mixtures of volatile constituents at temperature environment, most of which originate from secondary metabolism produced and stored in their own secretory structures formed in the leaves, flowers, branches/stems or roots of various species (Kamatou et al. 2007). In various plant families, namely Lamiaceae and Verbenaceae, usually secreted by glandular trichomes, which have various forms, structures and functions distributed mainly on the surface of the leaves (Kamatou et al. 2007). As specialized cell structures, trichomes constitutes specialized organs for EO synthesis and protection, protect the plant from its own toxicity and certain constituents, such as monoterpenes, which concentrations are phytotoxic. Lamiaceae trichomes presents two morphologically distinct structures: naked and pilous, distinguished by their structure and form of secretion, a fact evidenced by microscopic observations optics and electronics (Hallahan 2000).

Table 1. Pharmacological activities of isolated plant compounds

Name	Species	Biologic Function	References
Thionins	*Arabidopsis thaliana* *Pyrularia pubera* *Crambe abyssinica*	AB, CT, AF, AC	(Dang, Van Damme, 2015; Guzmán-Rodríguez et al. 2015)
Defencins	*Vigna unguiculata* *Phaseolus lunatus* *P. vulgaris* *P. angularis*	AB, CT, AF, AC	(Cândido et al. 2014, Guzmán-Rodríguez et al. 2015)
Cyclotides	*Viola abyssinica* *V. odorata* *Oldenlandia affini* *Clitoria ternatea*	AB, AV, CT, AC	(Cândido et al. 2014, Guzmán-Rodríguez et al. 2015)

Antibacterial (AB), Antiviral (AV), Antifungal (AF), Anticancer (AC) and Cytotoxic (CT) activities.

Those activities are intended to protect the plant, however, since early humanity has discovered that having diverse properties in the plant and against its predators and parasites, these could also be of great utility in the maintenance and restoration of human health. Thus, over time, we have studied the potential of medicinal and aromatic plants. Potentials are achieved because of the chemical nature and the percentage of constituents of their essential oils (Cunha et al. 2012). The main activities studied are antimicrobial activity, namely antibacterial, antifungal and antiviral, anxiolytic, antidepressive, anti-inflammatory, antioxidant, anticarcinogenic and antinociceptive as summarized in Table 1.

The EOs were long considered to be "physiological waste" or products of detoxification, as it was said, moreover, of the products of metabolism secondary education in general. However, with the scientific advances, they are attributed to the EO ecological functions, such as inhibition of seed germination (allelopathy), protection against predators and attraction of pollinators (Santos 2000; Barreiro 2006). Several biological activities of the EO are recognized, inter alia, antimicrobial activity, while at the same time confirming its low toxicity in mammals and their low negative environmental impact. As a consequence, have been increasingly used in health care, resistance situations of pathogenic microorganisms to synthetic antibiotics and inhibiting the growth of microorganisms from contamination food and its toxins. EO applications include processing and preservation of food, pharmaceuticals, alternative medicine and therapies (Hayouni et al. 2008).

The mean EOs content per plant is variable and often low (<1%). At the however, some EOs present potent biological activity, in most not due to a single component, but rather to the combined action of their various constituents. As a rule, its composition is very complex, being frequently the presence of more than fifty constituents, usually identified by chromatographic and spectral techniques as terpenic compounds and/or phenylpropanoids.

The identification of essential oil constituents is important for the understanding and prediction of their physiological effects. For example, EO rich in sesquiterpene hydrocarbons often anti-inflammatory effect (Tappin et al. 2004) and, EO rich in trans-caryophyllene can lipo-regulation (Yoo et al. 2005). The terpenic constituents may be hydrocarbons (molecules consisting of carbon atoms and atoms of hydrogen, only) or may be ternary molecules containing oxygen atoms, in addition to carbon atoms and hydrogen atoms forming part of groups various functional groups, namely, alcohols, aldehydes, cyclic aldehydes, ketones, phenols, ethers, phenolic ethers, oxides or esters (Pengelly 2004).

There are essential oils capable of inhibiting the growth of Gram-positive (Gram+) and Gram-negative (Gram-) bacteria as well as fungi. The mode of action of the oils with active antimicrobial activity is associated with the presence of certain compounds having the ability to alter the permeability of the outer membrane of microorganisms and/or inhibit enzymes important for their growth and survival (Cunha et al. 2012).

MAIN SPECIES OF PLANTS USED TO OBTAIN ESSENTIAL OILS

Plants express a variety of bioactive compounds present in Eos and others products that provide resistance against possible injury caused by micro-organisms and/or macro-organisms. Some well-known families of proteins include: 2S albumins, glycine-rich proteins, vicilins, patatin, octatin, tarine, lectin, ribosome inactivating proteases (RIPs), protease and α-amylase inhibitors, ureases, arcellins and antimicrobial peptides (MPAs). Most of these proteins tend to accumulate in the vulnerable parts of the plant, such as seeds, nuts and grains; stem parenchyma; grains and legumes; and some roots and tubers. These organs are responsible for the synthesis and storage of proteins, presenting high protein content.

The storage proteins constitute an excellent source of amino acids and can be mobilized and used for plant maintenance, defense and growth, as well as in the embryonic and developmental stages (Carvalho and Guimarães 1981; Osborni et al. 1988; Munoz 1998; Olsnes 2004). Studies on plant compounds show that their harmful effects are involved in the defense of plants against other organisms with pathogens of fungal, viral and bacterial origin. In the context of storage proteins, some of these proteins have the characteristics of antimicrobial agents, acting in the defense of plants, and are also related to defense against unrelated pathogens, such as human pathogens (Cândido et al. 2011).

Plants are an important source of several molecules with pharmacological potential. They produce proteins, peptides and EOs as a natural defense mechanism, which can be expressed constitutively or induced in response to a pathogen attack on all organs. They are abundantly transcribed by a single gene and thus require less biomass and energy consumption in different plant species, and may represent up to 3% of the plant gene repertoire. Efficiency depends on several characteristics of the protein or peptide, including molecular mass, sequence, charge, conformation, secondary and tertiary structures, presence or absence of disulfide bonds and hydrophobicity (Wright and Wright 2005; Wright et al. 2007).

The major biological activities of proteins and EOs produced by plants are antifungal, antibacterial and herbivorous insects. In addition, they also exhibit enzymatic inhibitory activities and have roles in heavy metal tolerance, abiotic stress and development. With some showing cytotoxic activity against mammalian cells and/or anticancer activity against cancer cells of different origins. In plants, we can highlight three families of molecules of protein origin that contain members with cytotoxic and anticancer properties, defensins, thionines and cytoids (Stotz, Waller and Wang 2013).

The conventional view that proteins and EOs have an absolute structure and directly related to a single function goes against their ability to develop new functions. This expansive idea is also shown in the relationships between the protein and its structure, in which a single function for a protein is no longer so direct. Considering this logic, the idea of protein multifunctionality, in which multiple functions may be associated to a single peptide or protein structures, has been gaining attention in several fields of research (Nobeli, Favia and Thornton 2009; Franco 2011).

The denomination of multifunctionality that can be given for a certain object that performs several functions on its own. In the case of proteins and multifunctional peptides this denomination has been used a lot in recent years for proteins and peptides that alone play multiple functions, such as interactions with membranes of bacteria causing disruption and in the recruitment of macrophages, neutrophils an action defined as immunomodulation. These multiple functions are always associated with a single three-dimensional structure, with similar counterparts, and have gained attention in a number of fields of research, including immunology and enzymology (Franco et al. 2009).

Among principal plants species employed in EOs production we can describe *Melaleuca alternifolia* that exhibits antibacterial action on both Gram- as Gram+, including *Propionibacterium acnes*. The activity against this microorganism has allowed to be introduced into cosmetics for the treatment of acne skin with lesions inflammatory and non-inflammatory (Enshaieh et al. 2007). This oil has been shown to be multiresistant strains to conventional antibiotics, alternative in serious infectious situations (Warnke et al. 2013). This plant has presented good results in studies on its action against the Herpes simplex virus (Farag et al. 2004).

It is also very common to use this oil in the treatment of mycoses and other fungal infections, having demonstrated excellent action against *Candida albicans* (Ninomiya et al. 2012). It has special emphasis as antiseptic for oral affections. This one oil is currently widely used as topical antiseptic and in massages of relaxation (Cunha et al. 2013). In addition to the antimicrobial activities attributed to the essential oil of the tree "dochá," it is also possible to highlight its anti-inflammatory activity (Nogueira et al. 2014) and anxiolytic (Cunha et al. 2013).

Thymus vulgaris presents as composition of its EO thymol and p-cymene in higher concentrations and carvacrol, linalool, γ-terpinene, β-myrcene, geraniol, terpineol, terpinene-4-ol. These oils exhibit great polymorphism and can be found different chemotypes (Cunha et al. 2012) and proven effect under bacteria and fungi (Zuzarte et al. 2013, Rajkowska et al. 2014). This essential oil can be used in rheumatic pain, otitis, rhinitis, sinusitis and stomatitis, and also as antispasmodic, expelling, antioxidant (Cunha et al. 2012), anti-inflammatory and antiseptic (Zuzarte et al. 2013).

Rosmarinus officinalis L. belonging to the family Lamiaceae, is a sclerophyte plant with persistent leaf and well adapted to climate limitations (Papageorgiou et al. 2008), the flowers grow in inflorescences of the leaves present in the upper part of the branch (Vienna et al. 2005). Explored for the quality of its EO is used in traditional medicine as well as in as in the pharmaceutical, food and cosmetic industries, as raw material (Zaouali et al. 2005). The therapeutic properties of *Rosmarinus officinalis* have been known since for a long time, being a species traditionally used as a cholera agent and diuretic (Hoefler et al. 1987), anti-spasmodic, antiseptic, anti-depressant, anti-rheumatic, carminative, digestive and hypertensive (Kabouche et al. 2005). The chemical composition and antioxidant activity of rosemary oil has been mentioned in several publications. Although there are differences in EO chemical profiles, 1,8-cineole appears to be present in a significant amount in different populations of *R. officinalis* (Pintore et al. 2002; Gachkar et al. 2007). The characterization of EO from *R. officinalis* plants developed in Argentina, Portugal and Spain also revealed significant levels of verbenone, borneol or Myrcene (Lawrence 1997; Lorenzo et al. 1999).

Pelargonium graveolens L. is a very branched erect shrub that can reach a height of 1.3 m. The stem is herbaceous and covered with hair when the plant is young, becoming woody over time. It has leaves strongly cut and velvety texture due to the presence of glandular hairs, being a species of odor marked by the presence of citronellol and geraniol. It is widely used in Chinese homeopathy to promote the expulsion of toxins that inhibit the body's balance (Peterson et al. 2006). *Pelargonium* species are used in folk medicine for wound treatment, fever, colic, antibacterial and anthelmintic treatments (Lis-Balchin and Deans 1996). Methanolic extracts of *Pelargonium* species possess antibacterial and anti-oxidant activity (Lis-Balchin and Deans 1996). *P. graveolens* oil showed antifungal activity against six *Trichophyton* species (Shin and Lim 2004). In aromatherapy the essential oils of *Pelargonium* sp. are used in abnormal situations of menopause, nervous tension and anxiety (Rao 2002). The academic interest for the phytotherapeutic properties of *Pelargonium*, still little known or unexplored, follows with the popular and agroindustrial interest existing around several species of this genus (Lalli et al. 2008).

The essential oil of *Foeniculum vulgare* has a high trans-anethole content followed by components such as estragol, fenchone, α-pinene, limonene, myrcene, camphene, sabinene, β-mirene, β-pinene, α-phellandrene, and γ-terpinene (Özbek et al. 2003; Cunha et al. 2012). Its essential oil has spasmolytic activity on the smooth muscle, mucolytic, digestive and antiseptic (Cunha et al. 2012). It also shows antibacterial action on several species Gram + and Gram-, antifungal and antioxidant (Dadalioglu and Evrendilek 2004; Anwar et al. 2009). In addition, Özbek et al. (2003) has shown that this essential oil also has hepatoprotective action (Özbek et al. 2003). A recent publication by Oliveira et al. (2014) investigates the anti-ulcer power of several essential oil constituents that act on two main fronts: action against *Helicobacter pylori* and anti-inflammatory action on the gastrointestinal mucosa. The compounds having anti-ulcer properties are menthol, isopulegol, limonene, cineole, thymoquinone, carvacrol, α-terpineol, terpinen-4-ol, epoxycarvone, elemol, nerolidol, α-bisobolol, anethole, eugenol, 1'S1 acetochavicol and acetate of 1'S-1 Acetoxieugenol, cinnamaldehyde, cinnamic acid, citral, thymol and bisabolangelone (Oliveira et al. 2014).

Mentha piperita major compound is menthol, followed by menthol and other constituents such as cineole, menthyl acetate, isomentone, menthofuran, limonene, pulegone, isopulegol and carvone (Cunha et al. 2012). Since its major component, menthol, this essential oil presents spasmolytic, collagenous, antiseptic, carminative, anti-inflammatory and local anesthetic action (Sandberg and Corrigan, 2001). There are several publications confirming the interest of the use of this essential oil in the treatment of irritable colon syndrome due to its spasmolytic action. It is also quite frequent to use in affections of the respiratory tract like nasal congestion and cough associated to colds and flu, irritated throat and inflamed. It is also used in oral hygiene products as antiseptic and by the refreshing action. Precautions: This oil should not be administered in individuals with biliary obstruction, gall bladder inflammation and hepatitis (Cunha et al. 2012).

Santalum album has as main compounds Cis-α-santalol, α-santalal, cis-βsantalol are generally the compounds in higher concentration, presenting smaller amounts of α-curcumene, α-santalene, α-transbergamotene, β-curcumene, β-santalene, epi-β-santalene, santene, (z) -α-trans-bergamotol, bisabolol, lanceol, β-santalal (Burdock and Carabin 2008, Cunha et al. 2012). This essential oil has the ability to inhibit the growth of several Gram+ and Gram- strains (Figure 3), certain fungi and herpes simplex viruses (Sindhu et al. 2010). It presents high soothing/relaxing and anti-inflammatory power used in respiratory affections and inflammations of the mouth and pharynx (Sindhu et al. 2010, Cunha and Roque, 2013). It is useful in urinary tract infections (Burdock and Carabin 2008). Its antiproliferative effect of tumor cells has also been studied where it has demonstrated positive effects (Burdock and Carabin 2008).

The essential oil of *Syzygium aromaticum* has as main constituent's eugenol, eugenyl acetate and β-caryophyllene and antibacterial, antifungal and antiviral action (Fu et al. 2007, Naveed et al. 2013), being used in inflammations of the mouth and pharynx, dental caries and otitis. It is also associated with stimulation of gastric secretions and flatulence due to large amounts of eugenol (Cunha et al. 2012). This essential oil may be irritating to the mucous membranes in non-therapeutic doses. Ingestion is strongly discouraged during pregnancy and breast-feeding in children younger than six years or in patients with gastrointestinal problems or neurological disorders (Cunha et al. 2012).

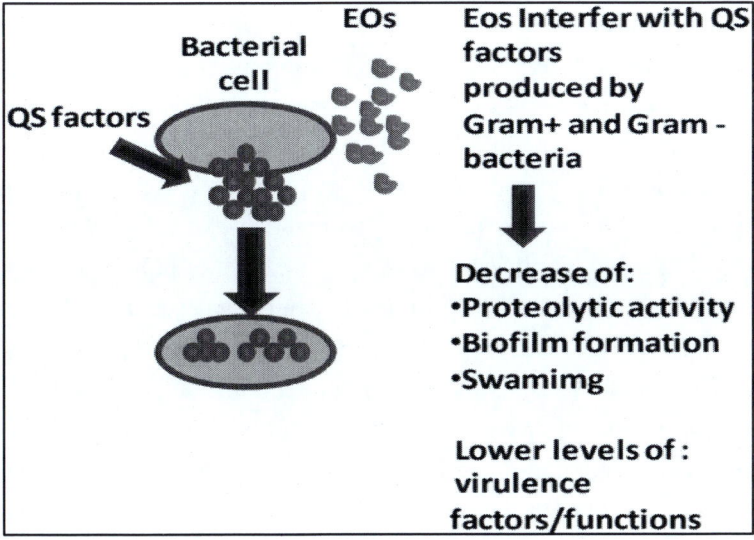

Figure 3. Mechanisms of anti-quorum sensing (QS) action by essential oils.

CONCLUSION

Essential oils are important and unexplored source of pharmacological molecules for clinical usage not only for its cosmetic application. It hides a great variety of compounds that can be used to apply major changes in health in the future. The use of these plant products has started from popular wisdom, since the beginning of human history, for the complementary therapies that marked the resurgence of these products. Nowadays, essential oils exhibit a high response to a variety of situations where people's health and well-being committed.

The associated risks with chemically synthesized drugs, adverse effects of various therapies, the existing treatments, and the imminent lack of curbing these diseases, have driven the search for new alternatives. In fact, research concerning essential oils is not that enough yet. However, because they are so complex in quantity and diversity of compounds, there are still large doubts and reluctance to use these products. Studies in this area have multiplied in the most varied health specialties, always being guided by empirical knowledge. Properties that are assigned by popular knowledge are exploited and many of these activities have been scientifically proven. But the step between research and application is not straightforward it takes a lot of time and work to correlate the results obtained *in vitro* and *in vivo* in animals with their use in safe use. It is therefore of great importance to continue to investigate and therapeutic properties of essential oils and to advance their application in the men, so that we can take advantage of everything that essential oils have to offer at the level of public health.

The exposure of plants to multiple environmental stressors, including predators and pathogens, gives these beings defense-to-survival strategies. Sice plants are sessile organisms, they are under constant pressure from the environment in which they are found. Within the co-evolutionary processes between plants and their predators - for example, fungi, bacteria, nematodes and insects - the improvement of resistance mechanisms occurs. It is the intrinsic ability to prevent, restrict or delay the penetration of a predator into the host tissue. These defenses play a critical role in regulating interactions between plants and herbivores, so understanding the evolution and ecology of plant defenses means understanding the origin and functioning of existing ecosystems, given that it is by the consumption of plant biomass - and therefore of overcoming the defensive barriers imposed against predation - that energy is always renewed, making the systems thermodynamically possible. In addition to the ecological appeal, the understanding of the barriers imposed by herbivory and the respective escapes developed by the herbivores allows a better management and development of cultivation strategies that are prone to avoid the loss of a considerable part of the agroindustrial production due to the lack of pest control.

The plants respond to the environment, in an excellent illustration of the "red queen theory," through various morphological, biochemical and molecular mechanisms to combat or mitigate the effects of pathogens attack - which are constantly dodging such mechanisms.

The biochemical defense mechanisms against pathogens are broad, highly dynamic and are mediated by responses that are either directly or indirectly, since the defensive compounds like Essential Oils can be produced in a constant and ubiquitous manner throughout the plant have many possibilities and an enormous potential to improve our clinical with new reservoir of molecules capable to give new treatments to well established human microbial diseases.

REFERENCES

AIA. 2014. "*Brief Story of Aromatherapy.*" Accessed July 19. http://www.alliance-aromatherapists.org/aromatherapy/brief-history-ofaromatherapy/.

Anwar, F., et al. 2009. "Antioxidant and antimicrobial activities of essential oil and extracts of fennel (*Foeniculum vulgare* Mill.) seeds from Pakistan." *Flavour and Fragance Journal.* 24:170-176. doi:/10.1002/ffj.1929.

Barreiro, A. P. 2006. "*Biomass production, yield and composition of basil essential oil (Ocimum basilicum L.) as a function of plant regulators.*" Masters dissertation, State University Paulista.

Benson, H. and Klipper, M. Z. 1977. "*Learning to relax.*" Rio de Janeiro: Artenova.

Burdock, George A and Ioana, G.Carabinb. 2008. "Safety assessment of sandalwool oil (*Santalum album* L.)." *Food and Chemical Toxicology.* 46:421-432. doi: 10.1016/j.fct.2007.09.092.

Burt, S. 2004. "Essential oils: their antibacterial properties and potential applications in foods—a review." *International Journal of Food Microbiology.* 94(3):223-53. doi: 10.1016/j.ijfoodmicro.2004.03.022.

Coelho, M. G. 2009. "*Essential oils for aromatherapy.*" Master Theses. University of Minho.

Corazza, S. 2002. "*Aromacology: a science of many smells.*" São Paulo: Senac.

Costanzo, L. S. 1999. "*Physiology.*" Rio de Janeiro: Guanabara Koogan.

Cunha, A. P. and Roque, O. R. 2013. "*Aromatherapy - Fundamentals and Use.*" Lisboa: Fundação Calouste Gulbenkian.

Cunha, A. P., et al. 2012. "*Aromatic Plants and Essential Oils Composition and Applications.*" Lisboa: Fundação Calouste Gulbenkian.

Dadalioglu, I. & Evrendilek, G. A. 2004. "Chemical compositions and antibacterial effects of essential oils of Turkish oregano (*Origanum minutiflorum*), bay laurel (*Laurus nobilis*), Spanish lavender (*Lavandula stoechas* L.), and fennel (*Foeniculum vulgare*) on common foodborne pathogens." *Journal of Agricultural and Food Chemistry.* 52: 8255-8260. doi: /10.1021/jf049033e.

Dang, Liuyi, and Van Damme, Els J. M. 2015. "Toxic proteins in plants." *Phytochemistry.* 117: 51–64.

Enshaieh, S., et al. 2007. "The efficacy of topical tea tree oil gel in mild to moderate acne vulgaris: A rondomized, double-blind placebo-controlled study Indian." *Journal of Dermatology Venereology and Leprology.* 73:22-25. doi: 10.4103/0378-6323.30646.

Farag, R. S., et al. 2004. "Chemical and Biological Evaluation of the Essential Oils of Different Melaleuca Species." *Phytotherapy Research.*18:30-35. doi: 10.1002/ptr.1348.

Gachkar, L. et al. 2007. "Chemical and biological characteristics of *Cuminum cyminum* and *Rosmarinus officinalis* essential oils." *Food Chemistry.* 102: 898-904. doi: 10.1016/j.foodchem.2006.06.035.

Guyton, Arthur C. M. D. et al. 1993. "*Basic Neuroscience: Anatomy and Physiology.*" Rio de Janeiro: Guanabara Koogan.

Hallahan, D. L. 2000. "Monoterpenoid biosynthesis in glandular trichomes of Labiate plants." *Advances in Botanical Research.* 31: 77-120. doi: 10.1016/S0065-2296(00)31007-2.

Hayouni, E. A. et al. 2008. "Tunisian *Salvia officinalis* L. and *Schinus molle* L. essential oils: their chemical compositions and their preservative effects against Salmonella inoculated in minced beef meat." *International Journal of Food Microbiology*. 125: 242-251. doi: 10.1016/j.ijfoodmicro.2008.04.005.

Hoare, J. 2010. *"Complete Aromatherapy Guide."* São Paulo: Editora Pensamento.

Hoefler, C. et al. 1987. "Comparative choleretic and hepatoprotective properties of young sprouts and total plant extracts of *Rosmarinus officinalis* in rats." *Journal of Ethnopharmacology*. 19(2): 133-143. doi: 10.1016/0378-8741(87)90037-7.

Johnson, C. B., et al. 2004. "Seasonal, Populational and Ontogenic Variation in the Volatile Oil Content and Composition of Individual of *Origanum vulgare* subsp. Hirtum, Assessed by GC Headspace Analysis and by SPME Sampling of Individual Oil Glands." *Phytochemical Analysis*. 15:286-292. doi:/10.1002/pca.780.

Kabouche, Z. 2005. "Comparative antibacterial activity of five Lamiaceae essential oils from Algeria." *Interantional Journal of Aromatherapy*. 15: 129-133. doi: 10.1016/j.ijat.2005.03.006.

Kamatou, G. P. P. et al. 2007. "Trichomes, essential oil composition and biological activities of *Salvia albicaulis* Benth. and *S. dolomitica* Codd, two species from the Cape region of South Africa." *South African Journal of Botany*. 73: 102-108. doi: 10.1016/j.sajb.2006.08. 001.

Lalli, J. Y. Y., Van Zyl, R. L., Van Vuuren, S. F., Viljoen, A. M. 2008. "In vitro biological activities of South African Pelargonium (Geraniaceae) species." *South African Journal of Botany*. 74: 153-157. doi: /10.1016/ j.sajb.2007.08.011.

Lawrence, B. M. 1997. "Progress in essential oils: rosemary oil." *Perfum. Flavor*. 22:1-71.

Lee, I. S. and Lee, G. J. 2006. "Effects of lavender aromatherapy on insomnia and depression in women college students." *Taehan Kanho Hakhoe*. 36:136-43. doi: 10.4040/jkan.2006.36.1.136.

Lis-Balchin, M. 2010. "Aromatherapy with essential oils." In: Baser, K. H. C. & Buchbauer, G. (eds.) *Handbook of essential oils - Science, Technology and Applications*. E.U.A: Taylor & Francis Group.

Lorenzo, D et al. 1997. "*Rosmarinus officinalis* L. (Labiatae) essential oils from the south of Brazil and Uruguay." *Journal of Essential Oil Research*. 11: 27-29. doi: 10.1080/10412905.1999.9701061.

Maluf, S. 2008. "*Aromatherapy: A Systemic Approach*." São Paulo: Samia Aromaterapia.

NAHA. 2014. "*Exploring Aromatherapie*." Accessed July 19. http://www.naha.org.

Nazzaro et al. 2013. "Effect of Essential Oils on Pathogenic Bacteria." *Pharmaceuticals*. 6:1451-1474. doi:10.3390/ph6121451.

Ninomiya, K., et al. 2012. "The Essential Oil of *Melaleuca alternifolia* (Tea Tree Oil) and Its Main Component, Terpinen-4-ol Protect Mice from Experimental Oral Candidiasis." *Biological and Pharmaceutical Bulletin*. 35:861-865. doi: 10.1248/bpb.35.861.

Nogueira, M. N. M., et al. 2014. "Terpinen-4-ol and alpha-terpineol (tea tree oil components) inhibit the production of IL-1b, IL-6 and IL-10 on human macrophages." *Inflammation Research*. 63:769-778. doi: 10.1007/s00011-014-0749-x.

Özbek, H., et al. 2003. "Hepatoprotective effect of *Foeniculum vulgare* essential oil." *Fitoterapia*. 74:317-319. doi: doi:10.1016/S0367-326X(03)00028-5.

Pintore, G. et al. 2002. "Chemical composition and antimicrobial activity of Rosmarinus officinalis L. oils from *Sardinia* and *Corsica*." *Flavour and Fragance Journal*. 17: 15–19. doi:10.1002/ffj.1022.

Rao, B. R. Rajeswara. 2002. "Biomass yield, essential oil yield and essential oil composition of rose-scented geranium (*Pelargonium* species) as influenced by row spacings and intercropping with cornmint (*Mentha arvensis* L. f. *piperascens* Malinv. ex Holmes)." *Industrial Crops and Products*. 16: 133-144. doi: 10.1016/S0926-6690(02)00038-9.

Santos, R. I. (2000) "*Pharmacognosy: from plant to medicine*." Santa Catarina, University of Rio Grande do Sul.

Serrano, A. Í. 1985. "*What is alternative medicine.*" São Paulo: Brasiliense.
Sibbritt, D. W. et al. 2014. "The self-prescribed use of aromatherapy oils by pregnant women." *Women and Birth*. 27:41-45. doi: 10.1016/j.wombi.2013.09.005.
Sindhu, R. K., et al. 2010. "*Santalum album* Linn: a review on morphology, phytochemistry and pharmacological aspects." *International Journal of PharmTech Research*. 2:914-919.
Tappin, M. R. R. 2004. "Quantitative chemical analysis for the standardization of copaiba oil by high resolution gas chromatograpy." *Química Nova*. 27(2): 236-240. doi: 10.1590/S0100-40422004 000200012.
Vera, M. N. and Vila, J. 2002. "Relaxation techniques." In: V. E. CABALLO. (eds), *Behavioral Modification and Therapy Techniques Manual*. São Paulo.
Warnke, P. H., et al. 2013. "The ongoing battle against multi-resistant strains: In vitro inhibition of hospital-acquired MRSA, VRE, Pseudomonas, ESBL *E. coli* and *Klebsiella* species in the presence of plant-derived antiseptic oils." *Journal of Cranio-MaxilloFacial Surgery*. 41:321-326. doi: 10.1248/bpb.35.861.
Yoo, S. S. et al. 2005. "The comparison of physicalchemical and textural properties, as well a volatile compound, from low-fat and regular-fat sausages." *International Journal of Food Science and Technology*. 40: 83-90. doi: 10.1111/j.1365-2621.2004.00911.x.
Zaouli, Y. et al. 2005. "Oil composition variability among populations in relationship with their ecological areas in Tunisian *Rosmarinus officinalis* L." *Flavour and Fragrance Journal*. 20: 512-520. doi:10. 1002/ffj.1428.
Zuzarte, M., et al. 2013. "Antifungal and anti-inflammatory potential of *Lavandula stoechas* and *Thymus herba-barona* essential oils." *Industrial Crops and Products*. 44:97-103. doi: 10.1016/j.indcrop.2012.11.002.

In: Antimicrobial Potential of Essential Oils   ISBN: 978-1-53616-945-4
Editors: B. Oliveira de Veras et al.   © 2020 Nova Science Publishers, Inc.

Chapter 3

MECHANISMS OF BACTERIAL RESISTANCE TO ANTIBIOTICS: ESSENTIAL OILS AS A STRATEGIC TOOL

Jalcinês da Costa Pereira, Jackelly Felipe de Oliveira, Rádamis Barbosa Castor, Lucas Silva Brito, Samuel de Souza Soares, Maria Vanessa Pontes da Costa Espínola, Fernanda Pereira Santos, Bruno Oliveira de Veras and Krystyna Gorlach-Lira**

Department of Molecular Biology, Center of Exact Sciences and Nature, Federal University of Paraíba, João Pessoa, Paraíba, Brazil

ABSTRACT

It is now well established that bacterial resistance to antibiotics has become a serious problem of public health that concerns almost all antibacterial agents and that manifests in all fields of their application. The increasing number of antibiotic resistant bacteria, through various mechanisms, such as efflux pumps and β-lactamase production, threatens

---

* Corresponding Author's Email: kglira@yahoo.com.

the world due to the possible emergence of highly pathogenic bacteria resistant to treatments using conventional antibiotics. Based on this perspective, a large number of studies are currently being conducted in search of new alternative drugs and tools to combat resistant bacterial pathogens. Essential oils enter this scenario as an alternative option to antimicrobial resistance, as they present remarkable antimicrobial action acting exclusively or in combination with antibiotics, increasing the susceptibility of these pathogens to drugs currently in use. Based on this perspective, the present work was elaborated in a review focused on two mechanisms of bacterial resistance to antibiotics (efflux pump and β-lactamase production), and the possibility of applying essential oils alone and in combination of antibiotics to overcome these mechanisms.

INTRODUCTION

The indiscriminate and exacerbated use of antimicrobial drugs has become frequent in everyday life, promoting the selection of bacterial strains resistant to these drugs (Knapp et al. 2010). Bacteria have developed sophisticated resistance mechanisms against antimicrobial drugs through specific responses such as mutational adaptations, acquisition of genetic material, or even alterations in gene expression, leading to resistance to most commonly used antibiotics in clinical practice (Munita et al. 2016). Bacterial resistance is a natural defense mechanism, of which there are several modes of action that the bacteria can use in order to be able to stop the action of antimicrobials drugs such as efflux pumps and production od enzymes such as β-lactamase, creating a resistance against this class of drugs, amongst other mechanisms (Kunishima et al. 2019).

Essential oils (EOs) are volatile secondary metabolites, responsible for the survival and perpetuation of plant species against various types of aggregations, such as attacks on pests, herbivores and phytopathogens. EOs have a complex composition, consisting of several classes of chemical constituents, varying between species, and abiotic factors to which plants are subjected (Stringaro et al. 2018; Wińska et al. 2019). Because of the complex composition of EOs, and their lipophilic nature, which allows compounds to pass through the cell membrane acting at various sites of microorganisms, EOs can be considered as an alternative for prospecting for new antimicrobials (Devi et al. 2010). There are several reports that show that EOs have as their main mechanism to act on the cellular surface, causing instability and thus leading to cellular death, and this effect could have a significantly stronger impact on gram-positive bacteria due to their lack of an extra membrane in comparison to gram-negative bacteria, which cause a larger diffusibility of lipophilic compounds (Hyldgaard et al. 2012; Bajpai et al. 2013).

Several biological activities of EOs are known as antibacterial (Thielmann et al. 2019), antifungal (Wang et al. 2018), antivirals (Gavanji et al. 2015), and molluscicide (Teixeira et al. 2012), moving billions of dollars annually, being a high economic value, which should reach the level of 11 billion dollars by 2022 worldwide (Zhai et al., 2018), these numbers coming from about 300 Eos used in the food, pharmaceutical, sanitary, cosmetic and perfumery industries (Tariq et al. 2019).

Based on this, this chapter has focused on bringing a literature review on antimicrobial resistance, with a major focus on efflux pump mechanisms and β-lactamases production, as well as demonstrating the potential of EOs as a strategic source to overcome bacterial resistance.

BACTERIAL RESISTANCE MECHANISMS TO ANTIBIOTICS

β-Lactamases Production

The production of β-lactamase is one of the most successful and important bacterial strategies among the mechanisms of resistance against β-lactam drugs, which are an important family of antibiotics. These enzymes catalyze the hydrolysis of the lactamic ring amide bond, rendering the antimicrobial ineffective (Munita et al. 2016). However, the resistance can also be motivated by the hyperexpression of efflux systems, alterations in membrane permeability as well as synthesis of penicillin-binding proteins with low affinity towards β-lactam antibiotics. In gram-negative bacteria such as *Pseudomonas aeruginosa*, that are resistant to several antibiotics, all of these mechanisms can exist simultaneously (Santos et al. 2015).

β-lactamases were described for the first time in the 40s (D'Costa et al. 2011). Once they have several enzymatic activities, β-lactamases have undergone several categorization attempts based on their biochemical attributes and in 1980, a first classification was proposed by Ambler. In this classification β-lactamases were divided according to their primary structures into four classes (A, B, C and D). According to differences in their catalytic mechanisms, these classes can still be classified into two groups: serine-β-lactamases (classes A, C and D) and metallo-β-lactamases (class B) (Ambler 1980; Bush 1995). More than thirty years after its creation, this classification still remains relevant.

Proposed in 1995, the functional classification of Bush-Jacob-Medeiros groups β-lactamases according to their functional similarities. This classification included three large groups, defined according to their substrate's inhibitory profiles, molecular mass and isoelectric point. In 2010, this classification was updated and then, molecular and functional characteristics were considered too, making it more comprehensive. New peptide sequences were added to the proposed list of requirements that describe new β-lactamases and now, there are at least 17 groups associated to four molecular classes (Bush et al. 2010).

The antimicrobial action of a β-lactam grug occurs by inhibiting the growth of bacteria by the acylation of a serine from the active site of Penicillin Binding Proteins (PBPs) and in the course of final steps of cell wall biosynthesis, cross-link peptide chains to form peptidoglycans, thereby leading to cell death (Spratt 1893; Bush et al. 2018).

The main resistance mechanisms used by gram-positive bacteria are different from those used by gram-negative. Gram-negative bacteria express enzymes that hydrolytically destroy the β-lactam antibiotics while gram-positive ones modify the target by the PBPs, altering their affinities for β-lactams while maintaining its physiological function (Ogawara 2015). However, early reports of β-lactam resistance in gram-positive bacteria date back to 1989, when after 20 years of penicillin use, clinical isolates of penicillin-resistant *Streptococcus pneumoniae* and

Staphylococcus aureus, produced PBPs variants of low affinity. This led to the appearance of penicillinase - β-lactamases that hydrolyze penicillin, and this new mechanism of resistance initially observed in *S. aureus*, challenged the therapeutic use of this agent.

Detailed studies based on genetic data show that most of the genes encoding β-lactamases are located in mobile DNA regions, such as plasmids, which further promotes the spreading of resistance among bacteria within the same and even different species.

β-lactamases included in class A (in Ambler's classification) have been described as a family that has a high level of hydrolytic activity, mainly against cefotaxime and ceftriaxone (Shibata et al. 2006), being the first enzyme capable of hydrolyzing a wide spectrum of cephalosporins at a clinically meaningful level. According to Davies et al. (2010), its genes and subsequent variants are a worldwide threat. Another very important and very widely studied family of enzymes are extended spectrum β-lactamases (ESBLs), which also belong to the class A of Bush's functional group 2b, their 6 types of enzymes, despite sharing similar hydrolytic profiles, from the genetic standpoint, are only remotely related. The enzymatic activity of ESBLs provides resistance not only to carboxypenicillins, but also to broad-spectrum cephalosporins, their hydrolytic activity can be amplified, with activity on carbapenems (Strateva et al. 2009). ESBLs are commonly found in enterobacteria, however, they have been increasingly found in other groups.

The enzymes metallo-β-lactamases (MBLs) belong to Ambler's class B, and class 3 of Bush-Jacoby-Medeiros' and usually use one or two of Zn^{2+} cations as a cofactor in the active site to facilitate β-lactam cleavage. MBLs are found worldwide in transmissible plasmids of numerous bacterial species. More than 80 different MBLs have been identified worldwide, where more than 75% of them are plasmid-encoded enzymes (Bush et al. 2010). They are broad-spectrum β-lactamases, active on practically all the β-lactam antibiotics used clinically. They therefore represent a serious clinical threat because of their ability to degrade carbapenems that are not degradable by most serine-β-lactamases (classes A, C and D) (Page 1999). MBLs are not sensitive to clinically available β-lactamase inhibitors (Drawz et al. 2010). According to Zhang et al. (2011) this is due to the lack of a clinically useful inhibitor and the main reason for that are the various active site structures, metal content and metal coordination residues.

Based on the diversity of metal binding, the MBLs were divided into 3 subclasses: B1, B2 and B3. Those belonging to subclass B1, including *Bacillus cereus* BcII (Hussainh et al. 1985), CcrA from *Bacteroides fragilis* (Rasmussen et al., 1990), IMP-1 from *Serratia marcescens* (Osano et al. 1994) and VIM -2 of *Pseudomonas aeruginosa* (Poirel et al. 2000), can bind up to two of Zn^{2+} at their active sites. They are active in mono- and dizinc forms, although the binding of the second zinc ion improves the activity, such as in BcII (Davies et al. 1974). In contrast, those belonging to subclass B2, including CphA (Massidda et al. 1991) and Imis (Walsh et al. 1996), are active only in the monozinc form. The binding of the second zinc ion inhibits the hydrolysis activity of CphA (Hernandez et al. 1997). MBLs belonging to subclass B3, such as L1 from *Stenotrophomonas maltophilia* (Felici et al., 1993), are active only in the dizinc form. The diversity of the active sites of MBLs makes it difficult to deduce a common mechanism of hydrolysis that may facilitate the design of mechanism-based inhibitors.

β-lactamases Class C (in Ambler's classification) account for much of the inherent resistance to chromosome-encoded β-lactams, and are mostly related to third generation cephalosporins. Livermore (2002) explains that the level of resistance depends on the degree of production of this enzyme, however, in *Pseudomonas aeruginosa* this mechanism confers resistance to β-lactams, even broad spectrum cephalosporins, except for carbapenemics.

β-lactamases Class D (in Ambler's classification) have different characteristics, but they are structurally related and exhibit excellent hydrolytic activity against oxacillin and similar compounds. Most extended spectrum oxacillinases are encoded by genes (*OXA-18 OXA-2, OXA-10* and *OXA-21*) located in plasmids or integrons, which contributes to their spread. The main clinical importance of this class is due to its potential activity against carbapenemics and broad spectrum cephalosporins.

According to Drawz et al. (2010), one way of reacting bacterial resistance to beta-lactams is the co-administration of antibiotics (lactamcs), with β-lactamase inhibitors. Serine-β-lactamase inhibitors are already available and in clinical use, such as clavulanate, sulbactam and tazobactam, however, due to the diversity of the active site and the lack of knowledge of the hydrolysis mechanism, the inhibitor development of these inhibitors has been hampered.

Efflux Pump

In recent decades, it was found that during the treatment of patients infected with one or more clinical isolates, these isolate that were initially susceptible to different classes of antibiotics, became less susceptible to many different molecules not related structurally (Li and Nikaido, 2009; Piddock, 2006).

Efflux pumps are transport proteins involved in the extrusion of toxic substrates (including virtually all classes of clinically relevant antibiotics) from cells into the external environment (Li and Nikaido 2009; Piddock 2006). It was previously reported that the balance in cell membrane permeability through the expression of efflux pumps efficiently controls the intracellular concentration of antibacterial agents (Nikaido and Pagès 2012).

These proteins are found in gram-negative and negative bacteria and may have specificity for a substrate or differentiated compounds (including antibiotics of various classes). Transporters can be divided in two major families, based upon bioenergetical and structural features, (i) primary transporters which hydrolyze ATP as a source of energy; they are also called ATP-binding cassette (ABC) transporters and, (ii) secondary transporters which utilize the proton (or sodium) gradient as a source of energy. In bacteria, secondary transporters are dominant and a number of them are MDR transporters.

ABC transporters are ubiquitous membrane systems, involved in different transport functions such as the efflux of toxins, metabolites and drugs. One of the most studied ABC transporter is the mammalian P-glycoprotein (P-gp, MDR1) whose overexpression confers resistance to cytotoxic compounds used in chemotherapy. Secondary transporters are classified into four families: the major facilitator superfamily (MFS), the resistance nodulation and cell division (RND) family, the small multi-drug resistance (SMR) family and the multidrug and toxic compound extrusion (MATE) family. The MFS proteins are ubiquitous systems, ensuring transport of sugars, intermediate metabolites and drugs. These proteins form two separate clusters, with either 12- or 14-transmembrane segments (TMS). The RND family is involved in the transport of lipophilic or amphiphilic molecules or toxic divalent cations; it is also responsible for the solvent tolerance of some bacterial strains. Membrane proteins of the SMR family are involved in the efflux of lipophilic cationic drugs in bacteria. They are the smallest drug efflux proteins known, with only 4 predicted TMS. They may function either as homo- or heterooligomeric complexes (Chopra et al. 2001; McDermott et al. 2003; Poole 2005).

To restore the effectiveness of antibiotics already used in clinical practice, efforts have been made to find inhibitors of these efflux systems. In this approach, the antibiotic is co-administered with a resistance-neutralizing inhibitor and therefore the antibiotic is still useful even against resistant organisms. This strategy can be used when resistance involves antibiotic inactivating enzymes and has been similar to adjuvants administered in conjunction with β-lactamase inhibitors, however, many adjuvants have been unable to be applied due to toxicity or already ineffective systems. efflux, and the constant search for new molecules of this capacity is required (Putman et al. 2000; Piddock et al. 2006; Poole 2007; Coutinho et al. 2010).

ESSENTIAL OILS: NEW ALTERNATIVE IN MICROBIAL CONTROL

Essential Oils

Essential oils are volatile compounds, synthesized by the secondary metabolism of plants, which play important roles in the defense against microorganisms, insects and herbivores. These compounds are commonly produced in flowers, leaves, stems, branches, seeds, fruits and roots, and are stored in secretory structures, cavities and channels.

EOs can be obtained by several methods, including: steam distillation, hydrodistillation, dry distillation and non-heat mechanical processes (Faleiro and Miguel 2013). The obtained chemical profile (number of molecules and stereochemistry) can differ according to the extraction method used. In addition, they may vary in quality, quantity and composition according to climate, soil composition, plant region, age and plant cycle (Bakkali et al. 2008).

The chemical composition is variable, however, among all components two or three are in higher concentration (20-70%), called main compounds, which determine the biological properties of the oil, while the other components are found in lower concentration. They are also responsible for performing activities through associations with the other components in reduced content. The components of essential oils have low molecular weight and belong to two groups of distinct biological origins, the main group being terpenes and the second one consisting of aromatic and aliphatic compounds (Yap et al. 2014).

Terpenes comprise a structurally and functionally diverse group formed by the combination of several units called isoprene (C5). Monoterpenes (C10) and sesquiterpenes (C15) are the main representatives of this group, however, it is possible to find other compounds with higher molecular mass, such as diterpenes (C20), triterpenes (C30) and tetraterpenes (C40). Monoterpenes make up 90% of the essential oils and can belong to several functional groups: hydrocarbons (e.g., mircene, p-cymene and pinenes), alcohols (e.g., geraniol, linalol and menthol), aldehydes (e.g., geranium, neral and citronellal), ketones (eg., tegetone, mentone, carvone and camphor), esters (e.g., propionate, citronellyl acetate and menthyl), ethers (e.g., 1,8-cineol and mentofuran), peroxides (e.g., ascaridol) and phenolic compounds (e.g., thymol and carvacrol) (Bakkali et al. 2008).

Essential Oils Overcoming Resistance of B-Lactamases

Unlike conventional antibiotics, essential oils have in their composition several chemical groups, of which they can act in various strucures of the microorganisms, presenting a multicomponent nature (Kon and Rai, 2012; Hammer et al., 2012; Oliva et al., 2018).

Various studies have shown the antibacterial potential of essential oils alone or in association with antibiotics used in clinical practice, as verified by Yap et al. (2014) who evaluated the antibacterial action of cinnamon (*Cinnamomum verum*) essential oil, alone and in combination with the antibiotic piperacillin, from which it was verified antibacterial activity of the essential oil, and synergistic effect with the antibiotic against *Escherichia coli* J53 MDR. Their results showed the potential to reverse the resistance of *E. coli* to piperacillin in two ways: modification of outer membrane permeability due to irreversible membrane damage or inhibition of bacterial quorum sensitivity (QS) by *C. verum* essential oil.

Moussoui and Alaoui (2016) studied the antimicrobial activity of essential oils from *Origanum compactum*, *Chrysanthemum coronarium*, *Thymus willdenowii*, *Melissa officinalis* and *Origanum majorana* alone and in combination with gentamicin, tobramycin, imipenem and ticarcillin, and observed that the essential oils act on some cases in synergy with antibiotics. The EO of *C. coronarium* exerted a synergistic effect against *K. pneumoniae*, *P. mirabilis*, *P. putida*, *S. aureus*, *E. aerogenes* and *E. coli*, and an antagonistic effect against *P. aeruginosa* and *S. enteritidis* when combined. with tobramycin. The EO of *T. willdenowii* in combination with ticarcillin, imipenem, gentamicin and tobramycin potentiated antibiotic activity. The combined use of EO of *O. majorana* with antibiotics showed higher antimicrobial activity against all bacterial strains analyzed.

A. baumannii is one of the most threatening pathogens on a global scale, particularly in health care facilities, due to its high resistance and increasing occurrence. Some studies show that synergy between essential oils and antibiotics offer an opportunity for the future development of treatment strategies for potentially lethal infections caused by *A. baumannii* (Alexic et al., 2014). Knezevic et al. (2016) studying the antibacterial action against *A. baumannii* strains by EO of *Eucalyptus camaldulensis* in combination with conventional antibiotics (ciprofloxacin and gentamycin) observed the synergistic action in most cases, still making some strains sensitive. Alexic et al. (2014) observed the synergistic action of essential oil of *Myrtus communis* L. in combination with antibiotic (ciprofloxacin) against *A. baumannii* strains isolated from the wounds. The use of essential oil promoted sensitivity of resistant strains, since MIC values were lower when the antibiotic was applied together with essential oil of *M. communis*.

Lahmar et al. (2017) found that the essential oils of *Pitanthus chloranthus*, *Teucruim ramosissimum* and *Pistacia lentiscus* in combination with the antibiotic's amoxicillin, tetracycline, piperacillin, ofloxacin and oxacillin were effective against resistant *S. aureus* and *A. baumannii* (MDRs).

Potent activity of essential oil of Tea Tree (*Malaleuca alternifolia*) against clinical strains of *S. aureus* MRSA, *K. pneumoniae* carbapenemic sensitive extended-spectrum beta-lactamase-producing pneumoniae (ESBL-CS-Kp), *K. pneumoniae* (CR-Kp), carbapenem resistant, have been reported by Oliva et al. (2018). The essential oil in combination with each reference antimicrobial showed a high level of synergism at sub-inhibitory concentrations, particularly with oxacillin (OXA) against MRSA. Yang et al. (2017) observed in *K. pneumoniae* (CR-Kp) a high antibacterial activity of cinnamon (*Cinnamomon verum*) essential oil, and in combination with the antibiotic meropenem this activity was potentiated, mediating the inhibition of beta-lactamase responsible for bacterial resistance.

Studies conducted with isolated components of essential oils have been presented as a new tool in antibiotic therapy, such as the study by Langeveld et al. (2014), who pointed out the possible synergistic effect of carvacrol, cinnamaldehyde, cinnamic acid, eugenol and thymol in combination with different classes of antibiotics. Palaniappan and Holley (2010) reported that thymol and carvacrol were effective in reducing the resistance of *Salmonella typhimurium* SGI 1(tetA). The synergistic effect between thymol, eugenol, berberine, cinnamaldehyde and streptomycin was also recorded against the food pathogens *Listeria monocytogenes* and *Salmonella typhimurium* (Liu et al. 2015).

These studies have shown that essential oils alone or in combination with antibiotics, have the ability to overcome bacterial resistance, which is by inactivation of their resistance mechanisms (inactivation of β-lactamases) by essential oils, appearing that they may become a promising source of adjuvant antibiotics or, in addition, to be used alone as antibiotic drugs.

Essential Oils Overcoming Resistance of Efflux Pumps

The synergistic action with natural products usually occurs through the inhibition of a resistance mechanism present in bacteria, the most common being efflux pumps. Several studies have demonstrated the effectiveness of essential oils in inactivating this mechanism, highlighting the bacterial resistance, as the study by Fadli et al. (2014) demonstrated that the natural compounds synthesized by *Thymus maroccanus* are capable of change of bacterial resistance associated with plasmatic membrane. These mechanisms, including efflux pumps, offer a first line of defense not only against natural molecules but also against the most common types of antibiotics, which is something that supports the presence of these resistance-expressing bacteria on environments, despite the lack of antibiotics in them.

Mouwakeh et al. (2018) showed how *Nigella sativa* essential oil and some of its active compounds (thymokinone, carvacrol and p-cymene) act against the pathogen *Listeria monocytogenes*. From the obtained results, they concluded that the oil and its components acted in the membrane integrity and in the efflux pumps present in this pathogen. This was probably due to reduced cellular excretion via efflux pumps, as well as reduced membrane integrity, promoting increased membrane permeability. Sewanu et al. (2015) also observed the same activity pattern of *Eucalyptus grandis* essential oil, which presented both mechanisms: efflux pump inhibition and plasma membrane modifications in *Klebsiella pneumoniae*, *Staphylococcus aureus* and *Moraxella catarrhalis*.

Another compound capable of acting on these defense mechanisms was essential oil of *Rosmarinus officinali*, popularly known as rosemary. The EO from this plant demonstrated synergic activity with ciprofloxacin, acting on the restoration of this antibiotic's properties against multiresistant pathogens such as *Acinetobacter baumannii* and *Pseudomonas aeruginosa*. The two mechanisms observed by Saviuc et al. (2016) were the increase in membrane permeability and the inhibition of efflux bombs, much like the findings of Mouwakeh et al. (2018) and Sewanu et al. (2015).

Several studies have been conducted to more specifically examine under which efflux pumps essential oils may be acting using strains that have genes encoding these transmembrane proteins. Coutinho et al. (2010) used a strain of *Staphylococcus aureus* SA 1199B that overexpresses the *norA* gene encoding norfloxacin efflux (NorA) proteins responsible for intracellular elimination of fluoroquinolones such as norfloxacin. They found that *Croton zehntneri* essential oil inhibited the efflux pump by increasing the fluoroquinolones cell concentration.

Chovanova et al. (2015), evaluated the performance of the essential oils of *Salvia fruticosa*, *Salvia officinalis* and *Salvia sclarea* against a tetracycline resistant clinical isolate *Staphylococcus epidermidis*, verifying that all EOs showed antibacterial activity and were able to act synergistically with tetracycline, by inhibition of efflux pumps and repressing its coding gene (Tet^r).

Furthermore, the literature brings us several studies that verify the action of other OEs against species of microrganisms that overexpress certain genes responsible for ensuring their resistance, such as is the case with efflux pumps. The EO of *Helichrysum italicum* is capable of modulating bacterial resistance through its activity on efflux pumps. Its inhibitory activity was assessed by Lorenzi et al. (2009) against the pathogens *Enterobacter aerogenes*, *E. coli*, *P. aeruginosa* and *Acinetobacter baumannii*. In this study, the *H. italicum* essential oil decreased the MIC values of chloramphenicol allied against all of the isolates, including one strain that superexpresses the tripartite efflux pump AcrAB-TolC (*E. aerogenes* CM-64). It was also capable of restoring susceptibility in one strain that overexpressed efflux pumps different from AcrAB, (*E. coli* AG100A Tetr).

Fadli et al. (2011), on the other hand, studied the modes of action of several species of *Marrocan* plants in order to identify the compounds that could inhibit efflux pumps in 5 different bacterial species (*Escherichia coli, Enterobacter aerogenes, Klebsiella pneumoniae, Salmonella enterica* serovar Typhimurium and *Pseudomonas aeruginosa*) and also some multiresistant strains. The EOs obtained from *Thymus maroccanus* and *Thymus broussonetii* increased the susceptibility to chloramphenicol in most of the isolates, especially *P. aeruginosa* PA01 and *P. aeruginosa* PA124 (super expresses the protein OprM) and also increased this drug's efficacy against lineages that superproduce the protein AcrAB (*E. coli* AG102) and another multiresistant lineage (*E. aerogenes* EA27).

CONCLUSION

In conclusion, the EO and its components have antibacterial activity against Gram-negative and positive bacteria, being this activity performed entirely by changes in the bacterial membrane, enzymatic inhibition (β-lactamases) and inactivation of efflux systems (efflux pumps). The synergy mechanisms of OEs with conventional antibiotics are very complex and not yet fully understood, as they involve multiple interactions between individual OEs constituents on the one hand and individual OEs constituents and antibiotics on the other. OEs generally have different antibacterial effects than their major components, suggesting synergistic, additive or antagonistic interactions between their constituents and also an undisputed contribution of minor components to antibacterial activity. The complexity of the interactions generated by combining OEs and antibiotics is evident. Overall, the results of *in vitro* tests are very promising for the development of new effective formulations against Gram-negative bacteria, and further *in vivo* studies are needed to assess the bioavailability, efficacy and toxicity of EOs/antibiotic combinations.

REFERENCES

Alexic, V., Mimica-Dukic, N., Simin, N., Nedeljkovic, N. S. & Knezevic, P. (2014). "Synergistic effect of *Myrtus communis* L. essential oils and conventional antibiotics against multidrug resistant *Acinetobacter baumannii* wound isolates". *Phytomedicine.*, *21*(12), 1666-1674. Doi: 10.1016/j.phymed.2014.08.013.

Ambler, R. P. (1980). "The structure of β-lactamases." *Philos. Trans. R. Soc. Lond. B Biol. Sci.*, *289*(36), 289:321–331. Doi:10.1098/rstb.1980.0049.

Bajpai, V. K., Sharma, A. & Baek, K. H. (2013). "Antibacterial mode of action of *Cudrania tricuspidata* fruit essential oil, affecting membrane permeability and surface characteristics of food-borne pathogens." *Food Control.*, *32*(2), 582-90. Doi: 10.1016/j.foodcont.2013.01.032.

Bakkali, F., Averbeck, S., Averbeck, D. & Idaomar, M. (2008). "Biological effects of essential oils – a review." *Food and Chemical Toxicology.*, *46*(2), 446-75. Doi: 10.1016/j.fct.2007.09.106.

Bush, K. (2018). "Past and present perspectives on β-lactamases." *Antimicrob. Agents Chemother.*, *62*(10), 01076-18. Doi: 10.1128/AAC.01076-18.

Bush, K., Jacobi, G. A. & Medeiros, A. A. (1995). "A functional classification scheme for β-lactamases and its correlation with molecular structure." *Antimicrob. Agents Chemother.*, *39*(6), 1211–1233. Doi:10.1128/AAC.39.6.1211.

Bush, K. & Jacoby, G. A. (2010). "Classificação funcional atualizada de β lactamases". *Antimicrob Agents Chemother.*, *54*, 969 – 976. Doi: 10.1128/AAC.01009-09.

Chopra, I. & Roberts, M. (2001). Tetracycline antibiotics: mode of action, applications, molecular biology, and epidemiology of bacterial resistance. *Microbiol. Mol. Biol. Rev.*, *65*, 232–60. Doi: 10.1128/MMBR.65.2.232-260.2001.

Chovanova, R., Mezovska, J., Vaverkova, S. & Mikulasova, M. (2015). "The inhibition the Tet(K) efflux pump of tetracycline resistant *Staphylococcus epidermidis* by essential oils from three *Salvia* species". *Letters in Applied Microbiology.*, *61*, 1, 58-62. Doi: 10.1111/lam.12424.

Coutinho, H. D. M., Matias, E. F. F., Santos, K. K. A., Tintino, S. R., Souza, C. E. S., Guedes, G. M. M., Santos, F. A. D., Costa, J. G. M. et al. (2010). "Enhancement of the norfloxacin antibiotic activity by gaseous contact with the essential oil of *Croton zehntneri*". *Journal of Young Pharmacists.*, *2*, (4), 362-364. Doi: 10.4103/0975-1483.71625.

Davies, R. B. & Abraham, E. P. (1974). "Metal cofactor requirements of beta-lactamase II". *Biochem. J.*, *143*(1), 129–135. Doi: 10.1042/bj1430129.

D'Costa, V. M., King, C. E., Karlan, L., Morar, M., Contado, W. W. L., Schwaez, C., Froese, D., Zazula, G., Calmels, F., Debruyne, R., Golding, G. B. & Hendrik, N. (2011). "Antibiotic resistance is ancient." *Nature.*, *477*, 457-461. Doi: 10.1038/nature10388.

Devi, K. Pandima., Nisha, S. Arif., Sakthivel, R. & Pandian, S. Karutha. (2010). "Eugenol (an essential oil of clove) acts as an antibacterial agent against *Salmonella typhi* by disrupting the cellular membrane." *J. Ethnopharmacol.*, *130*(1), 107-15. Doi:10.1016/j.jep.2010.04.025.

Drawz, S. M. & Bonomo, R. A. (2010). "Three decades of beta-lactamase inhibitors". *Clin. Microbiol. Rev.*, *23*(1): 160–201. Doi: 10.1128/CMR.00037-09.

Fadli, M., Chevalier, J., Hassani, L., Mezrioui, N. E. & Pages, J. M. (2014). "Natural extracts stimulate membrane-associated mechanisms of resistance in Gram-negative bacteria". *Lett. Appl. Microbiol.*, *58*, 472– 477. Doi: 10.1111/lam.12216.

Fadli, M., Chevalier, J., Saad, A., Mezrioui, N. E., Hassani, L. & Pages, J. M. (2011). "Moroccan plants as potential chemosensitisers restoring antibiotic activity in resistant Gram-negative bacteria". *International Journal of Antimicrobial Agents.*, *38* (4), 325-30. Doi:10.1016/ j.ijan timicag.2011.05.005.

Faleiro, M. L. & Miguel, M. G. (2013). "Use of essential oils and their components against multidrug-resistant bacteria." In: *Fighting Multidrug Resistance with Herbal Extracts, Essential Oils and Their Component.*, 65-94. Ed.: M. K. Rai and K.V. Kon. Academic Press. Doi: 10.1016/B978-0-12-398539-2.00006-9.

Felici, A., Amicosante, G., Oratore, A., Strom, R., Ledent, P., Joris, B., Fanuel, L. & Frere, J. M. (1993). "An overview of the kinetic parameters of class B beta-lactamases". *Biochem. J.*, *291*(1), 151–155. Doi: 10.1042/bj2910151.

Gavanji, S., Sayedipour, S. S., Larki, B. & Bakhtari, A. (2015). "Antiviral Activity of some plant oils against herpes simplex virus type 1 in vero cell culture." *Journal of Acute Medicine.*, *5*(3), 62-68. Doi:10.1016/ j.jacme.2015.07.001.

Hammer, K. A., Carson, C. F. & Riley, T. V. (2012). "Effects of *Melaleuca alternifolia* (tea tree) essential oil and the major monoterpene component terpinen-4-ol on the development of single- and multistep antibiotic resistance and antimicrobial susceptibility". *Antimicrob. Agents Chemother.*, *56*(2), 909–915. Doi:10.1128/ AAC.05741-11.

Hernandez, V. M., Felici, A., Weber, G., Adolph, H. W., Zeppezauer, M., Rossolini, G. M., Amicosante, G., Frere, J. M. & Galleni, M. (1997). "Zn(II) dependence of the *Aeromonas hydrophila* AE036 metallo-beta-lactamase activity and stability". *Biochemistry.*, *36*(38), 11534–11541. Doi:10.1021/bi971056h.

Hussain, M., Carlino, A., Madonna, M. J. & Lampen, J. O. (1985). "Cloning and sequencing of the metallothioprotein betalactamase II gene of *Bacillus cereus* 569/H in *Escherichia coli*." *J. Bacteriol.*, *164*(1), 223–229. PMCID: PMC214233.

Hyldgaard, M., Mygind, T. & Meyer, R. L. (2012). "Essential oils in food preservation: mode of action, synergies, and interactions with food matrix components." *Front. Microbiol.*, *3*, 12. Doi: 10.3389/ fmicb. 2012.00012.

Knapp, C. W., Dolfing, J., Ehlert, P. A. I. & Graham, D. W. (2010). "Evidence of increasing antibiotic resistance gene abundances in archived soils since 1940." *Environmental Science & Technology.*, *44*(2), 580-87. Doi:10.1021/es901221x.

Knezevic, P., Aleksic, V., Simin, N., Svircev, E., Petrovic, A. & Mimica-Dukic, N. (2016). "Antimicrobial activity of *Eucalyptus camaldulensis* essential oils and their interactions with conventional antimicrobial agents against multi-drug resistant *Acinetobacter baumannii*". *J. Ethnopharmacol.*, *178*, 125–136. Doi: 10.1016/j.jep.2015.12.008.

Kon, K. V. & Rai, M. K. (2012). "Plant essential oils and their constituents in coping with multidrug-resistant bacteria". *Expert. Rev. Anti. Infect. Ther.*, *10*(7), 775–790. Doi: 10.1586/eri.12.57.

Kunishima, H., Ishibashi, N., Wada, K., Oka, K., et al. (2019). The effect of gut microbiota and probiotic organisms on the properties of extended spectrum beta-lactamase producing and carbapenem resistant *Enterobacteriaceae* including growth, beta-lactamase activity and gene transmissibility. *J. Infect. Chemother.* pii: S1341-321X(19)30134-5. Doi: 10.1016/j.jiac.2019.04.021.

Lahmar, A., Bedoui, A., Mokdad-Bzeouich, I., Dhaouifi, Z., Kalboussi, Z., Cheraif, I., Ghedira, K. & Chekir-Ghedira, L. (2017). "Reversal of resistance in bacteria underlies synergistic effect of essential oils with conventional antibiotics". *Microb. Pathog.*, *106*, 50–59. Doi:10.1016/ j.micpath.2016.10.018.

Langeveld, W. T., Veldhuizen, E. J. A. & Burt, S. A. (2014). "Synergy between essential oil components and antibiotics: a review". *Critical Reviews in Microbiology.*, *40*(1), 76–94. Doi:10.3109/ 1040841X.201 3.763219.

Li, X. Z. & Nikaido, H. (2009). Efflux-mediated drug resistance in bacteria: an update. *Drugs.*, *69*(12), 1555–1623. Doi:10.2165/113170 30-000000000-00000.

Livermore, D. M. (2002). "Multiple mechanisms of antimicrobial resistance in Pseudomonas aeruginosa: our worst nightmare?" *Clin. Infect. Dis.*, *34*(5), 634-40. Doi: 10.1086/338782.

Liu, Q., Niu, H., Zhang, W., Mu, H., Sun, C. & Duan, J. (2015). "Synergy among thymol, eugenol, berberine, cinnamaldehyde and streptomycin against planktonic and biofilm-associated food-borne pathogens". *Lett. Appl. Microbiol*, *60*(5), 421–430. Doi: 10.1111/lam.12401.

Lorenzi, V., Muselli, A., Bernardini, A. F., Berti, L., Pagès, J. M., Amaral, L. & Bolla, J. M. (2009). "Geraniol restores antibiotic activities against multidrug-resistant isolates from gram-negative species". *Antimicrobial Agents and Chemotherapy.*, *53*(5), 2209-11. Doi:10.1128/AAC.00919-08.

Massidda, O., Rossolini, G. M. & Satta, G. (1991). "The Aeromonas hydrophila cphA gene: molecular heterogeneity among class B metallo-beta-lactamases". *J. Bacteriol.*, *173*(15), 4611–4617. Doi: 10.1128/jb.173.15.4611-4617.1991.

McDermott, P. F., Walker, R. D. & White, D. G. (2003). Antimicrobials: modes of action and mechanisms of resistance. *Int. J. Toxicol.*, *22*, 135–43. Doi:10.1080/10915810305089.

Mouwakeh, A., Telbisz, Á., Spengler, G., Mohácsi-Farkas, C. & Kiskó, G. (2018). "Antibacterial and resistance modifying activities of *Nigella sativa* essential oil and its active compounds against *Listeria monocytogenes*". *In Vivo.*, *32*(4), 737-743. Doi:10.21873/*invivo*.1 1302.

Moussaoui, F. & Alaoui, T. (2016). Evaluation of antibacterial activity and synergistic effect between antibiotic and the essential oils of some medicinal plants. *Asian Pac. J. Trop. Biomed.*, *6* (1), 32–37. Doi: 10.1016/j.sciaf.2019.e00090.

Munita, J. M. & Arias, C. A. (2016). "Mechanisms of Antibiotic Resistance". *Microbiol. Spectr.*, *4*(2), 02-37. Doi: 10.1128/microbiolspec.VMBF-0016-2015.

Nikaido, H. & Pagès, J. M. (2012). Broad-specificity efflux pumps and their role in multidrug resistance of Gram-negative bacteria. *FEMS Microbiol Rev.*, *36*(2), 340-63. Doi: 10.1111/j.1574-6976.2011. 00290.x.

Ogawara, H. (2015). "Penicillin-binding proteins in Actinobacteria." *The Journal of Antibiotics.*, *68*(4), 223–245. Doi: 10.1038/ja.2014.148.

Oliva, A., Costantini, S., De Angelis, M., Garzoli, S., Božović, M., Mascellino, M. T., Vullo, V. & Ragno, R. (2018). "High potency of *Melaleuca alternifolia essential* oil against multi-drug resistant gram-negative bacteria and methicillin-resistant *Staphylococcus aureus*". *Molecules.*, *23*(10), 2584. Doi:10.3390/molecules23102584.

Osano, E., Arakawa, Y., Wacharotayankun, R., Ohta, M., Horii, T., Ito, H., Yoshimura, M. & Kato, N. (1994). "Molecular characterization of an enterobacterial metallo beta-lactamase found in a clinical isolate of *Serratia marcescens* that shows imipenem resistance". *Antimicrob. Agents Chemother.*, *38*(1), 71-8. Doi:10.1128/aac.38.1.71.

Page, M. I. (1999). "The reactivity of beta-lactams, the mechanism of catalysis and the inhibition of beta-lactamases". *Curr. Pharm. Des.*, *5*, 895–913.

Palaniappan, K. & Holley, R. A. (2010). "Use of natural antimicrobials to increase antibiotic susceptibility of drug resistant bacteria". *Int. J. Food Microbiol.*, *140*(2-3), 164–168. Doi:10.1016/j.ijfoodmicro.2010.04.001.

Piddock, L. J. (2006). Multidrug-resistance efflux pumps—not just for resistance. *Nat. Rev. Microbiol.*, *4*, 629–36. Doi:10.1038/nrmicro1464.

Poirel, L., Naas, T., Nicolas, D., Collet, L., Bellais, S., Cavallo, J. D. & Nordmann, P. (2000). "Characterization of VIM-2, a carbapenem-hydrolyzing metallo-beta-lactamase and its plasmid- and integron-borne gene from a *Pseudomonas aeruginosa* clinical isolate in France". *Antimicrob. Agents Chemother.*, *44*(4), 891–897. Doi: 10.1128/aac.44.4.891-897.2000.

Poole, K. (2005). Efflux-mediated antimicrobial resistance. *J. Antimicrob. Chemother.*, *56*, 20–51. Doi:10.1093/jac/dki171.

Poole, K. (2007). Efflux pumps as antimicrobial resistance mechanisms. *Annals of Medicine.*, *39*(3), 162–176. Doi:10.1080/0785389070 1195262

Putman, M., van Veen, H. W. & Konings, W. N. (2000). Molecular properties of bacterial multidrug transporters. *Microbiol. Mol. Biol. Rev.*, *64*, 672–93. Doi:10.1128/mmbr.64.4.672-693.2000.

Rasmussen, B. A., Gluzman, Y. & Tally, F. P. (1990). "Cloning and sequencing of the class B beta-lactamase gene (ccrA) from *Bacteroides fragilis* TAL3636". *Antimicrob. Agents Chemother.*, *34*, 1590–1592. Doi: 10.1111/j.1365-2958.1991.tb01895.x.

Santos, I. A. L., Nogueira, J. M. R. & Mendonça, F. C. R. (2015). "Mecanismos de resistência antimicrobiana em *Pseudomonas aeruginosa*". *RBAC: Revista Brasileira de Análises Clínicas.*, *47*(1/2), 5-12. https://www.arca.fiocruz.br/handle/icict/15160.

Saviuc, C. M., Gheorghe, I., Coban, S., Drumea, V., Chifiriuc, M. C., Banu, O. A. M., Bezirtzoglou, E. E. V. & Lazăr, V. (2016). "*Rosmarinus officinalis* essential oil and eucalyptol act as efflux pumps inhibitors and increase ciprofloxacin efficiency against *Pseudomonas aeruginosa* and *Acinetobacter aumannii* MDR strains". *Romanian Biotechnological Letters.*, *21*(4), 11782-11790. https://e-repository.org/rbl/vol.21/iss.4/20.pdf.

Sewanu, S. O., Bongekile, M. C., Folusho, O. O., Adejumobi, L. O. & Rowland, O. A. (2015). "Antimicrobial and efflux pumps inhibitory activities of *Eucalyptus grandis* essential oil against respiratory tract infectious bacteria". *Journal of Medicinal Plants Research.*, *9* (10), 343-348. Doi:10.5897/JMPR2015.5652.

Shibata, N., Kurokawa, H., Doi, Y., Yagi, T., Yamane, K., Wachino, J., Set al. (2006). "PCR classification of CTX-M-type β-lactamase genes identified in clinically isolated Gram-negative bacilli in Japan". *Antimicrob. Agents Chemother.*, *50*(2), 791-5. Doi: 10.1128/ AAC.50.2.791-795.2006.

Spratt, B. G. (1983). "Penicillin-binding proteins and the future of ß-lactam antibiotics". *J. Gen. Microbiol.*, *129*, 1247–1260. Doi:10.1099/ 00221287-129-5-1247.

Strateva, T. & Yordanov, D. (2009). "*Pseudomonas aeruginosa* - a phenomenon of bacterial resistance". *J. Med. Microbiol.*, *58*(.9), 1133-48. Doi: 10.1099/jmm.0.009142-0.

Stringaro, A., Colone, M. & Angiolella, L. (2018). "Antioxidant, antifungal, antibiofilm, and cytotoxic activities of *Mentha* spp. essential oils." *Medicines (Basel)*, *5*(4), 112. Doi:10.3390/medicines5040112.

Tariq, S., Wani, S., Rasool, W., Shafi, K; Bhat, M. A., Prabhakar, A., Shalla, A. H. & Rather, M. A. (2019). "A comprehensive review of the antibacterial, antifungal and antiviral potential of essential oils and their chemical constituents against drug-resistant microbial pathogens." *Microb. Pathog.*, *134*. Doi:10.1016/ j.micpath.2019. 103580.

Teixeira, T., Rosa, J. S., Rainha, N., Baptista, J. & Rodrigues, A. (2012). "Assessment of molluscicidal activity of essential oils from five azorean plants against *Radix peregra* (Müller, 1774)." *Chemosphere.* 87(1): 1-6. Doi:10.1016/j.chemosphere.2011.11.027.

Thielmann, J., Muranyi, P. & Kazman, P. (2019). "Screening essential oils for their antimicrobial activities against the foodborne pathogenic bacteria *Escherichia coli* and *Staphylococcus aureus*." *Heliyon*, *5*(6), 1-6. Doi:10.1016/j.heliyon.2019.e01860.

Wang, H., Yang, Z., Ying, G., Yang, M., Nian, Y., Wei, F. & Kong, W. (2018). "Antifungal evaluation of plant essential oils and their major components against toxigenic fungi." *Industrial Crops and Products.*, *120*, 180-86. Doi:10.1016/j.indcrop.2018.04.053.

Walsh, T. R., Gamblin, S., Emery, D. C., MacGowan, A. P. & Bennett, P. M. (1996). "Enzyme kinetics and biochemical analysis of ImiS, the metallo-beta-lactamase from *Aeromonas sobria* 163a". *J. Antimicrob. Chemother.*, *37*(3), 423–431.Doi: 10.1093 / jac / 37.3.423.

Wińska, K., Mączka, W., Łyczko, J., Grabarczyk, M., Czubaszek, A. & Szumny, A. (2019). "Essential oils as antimicrobial agents—myth or real alternative?". *Molecules*, *24*(11), 1-26. Doi: 10.3390/ molecules24112130.

Yang, S. K., Yusoff, K., Mai, C. W., Lim, W. M., Yap, W. S., Lim, S. E. & Lai, K. S. (2017). "Additivity vs synergism: investigation of the additive interaction of cinnamon bark oil and meropenem in combinatory therapy". *Molecules.*, *22*(11), e1733. Doi: 10.3390/molecules22111733.

Yap, P. S. X., Yiap, B. C., Ping, H. C. & Lim, S. H. E. (2014). "Essential oils, a new horizon in combating bacterial antibiotic resistance." *The Open Microbiology Journal.*, *8*, 6-14. Doi: /10.2174/1874285801408010006.

Zhai, H., Liu, H., Wang, S., Wu, J. & Kluenter, A. M. (2018). "Potential of essential oils for poultry and pigs." *Animal Nutrition.*, *4*(2), 179-86. Doi:10.1016/j.aninu.2018.01.005.

Zhang, H. M. & Hao, Q. (2011). "Crystal structure of NDM-1 reveals a common-lactam hydrolysis mechanism". *FASEB J.*, *25*(8), 2574–2582. Doi: 10.1096/fj.11-184036.

In: Antimicrobial Potential of Essential Oils ISBN: 978-1-53616-945-4
Editors: B. Oliveira de Veras et al. © 2020 Nova Science Publishers, Inc.

Chapter 4

ESSENTIAL OIL OF THE TEA TREE (*MELALEUCA ALTERNIFOLIA*): A POTENT ANTIMICROBIAL

Krystyna Gorlach-Lira*,
*Maria Vanessa Pontes da Costa Espínola,
Lucas Silva Brito, Fernanda Pereira Santos,
Jackelly Felipe de Oliveira, Samuel de Souza Soares,
Rádamis Barbosa Castor, Jalcinês da Costa Pereira
and Bruno Veras de Oliveira*
Department of Molecular Biology, Center of Exact Sciences
and Nature, Federal University of Paraíba, João Pessoa, Paraíba, Brazil

ABSTRACT

Tea Tree (*Melaleuca alternifolia*) is a medicinal aromatic plant found in tropical and subtropical regions and known for its antimicrobial properties. The essential oil of Tea Tree (TTO) is a complex mixture of terpenes, mainly mono and sesquiterpenes, of which terpinen-4-ol is the most representative. A number of studies have shown the effectiveness of

---

* Corresponding Author's Email: kglira@yahoo.com.

TTO in combating human and animal pathogenic microorganisms, and even those resistant to antibiotics such as methicillin-resistant *Staphylococcus aureus* (MRSA) and fluconazole-resistant *Candida albicans*. But despite widely investigated antimicrobial activity, the mechanisms of action by which TTO works remain uncertain. There are evidences that TTO, acts as a causative agent of disruption of the fungal and bacterial cell wall and plasma membrane, with consequent loss of metabolites and cell death. This revision focus on essential oil of *M. alternifolia* as the antibacterial and antifungal agent to be important alternative to conventional methods employed in the treatment of microbial infections, as well as brings insight to its mechanism of action against bacteria and fungi.

INTRODUCTION

Since ancient times, plants have been used as a medical treatment for various diseases and much of their therapeutic use comes from secondary metabolites, which are not vital for the direct survival of plants. The presence of these components allows plants to have advantages in their own perpetuation, such as self-defense functions and pollinator attraction, which are the main effects of essential oils (EOs) (Simões et al. 2017; Bhargava et al. 2013).

Essential oils are aromatic and highly volatile substances present in specialized glands of certain plants and their composition vary between the species of origin and abiotic conditions, being found terpenes, terpenoids, aromatic and aliphatic acids, amongst others (Ehlert et al. 2013; Sanli and Karadoğan 2016). Due to their complex versatility, essential oils can exert diverse actions towards pathogenic organisms using various mechanisms of action thst mostly are not clear yet. Nonetheless, due to their lipophilic nature, they can easily pass through the cytoplasmic membrane of microorganisms and act in the inhibition of enzymes that are crucial to their survival (Bajpai, Baek and Kang 2012).

Melaleuca alternifolia (tea tree) leaves are used to extract an essential oil of strong aroma, rich in mono and sesquiterpenes, with antioxidant, antifungal and antimicrobial activities (Li et al. 2016; Xu et al. 2017a). Many studies describe its use against various pathogenic microorganisms and elucidate the modes of action The essential oil of this plant started to be used officially since 1922 for therapeutic purposes, and became widespread in cosmetics, food, agriculture and other industries (Zhang et al. 2018).

The bioprospection of bioactive products has been intensifying due to the ever-growing microorganisms' resistance to the usual antimicrobial compounds, besides that, there's the increasing demand for natural compounds in pharmaceutical formulations because they are ecologically sustainable (Thosar et al. 2013). This review aimed to bring an overview of current knowledge on the components of tea tree essential oil and its antimicrobial activity, as well as to discuss the modes of action of TTO against bacteria and fungi, including multidrug-resistant microorganisms.

TEA TREE ESSENTIAL OIL COMPONENTS

The genus *Melaleuca* belongs to the Leptospermoideae subfamily of the Myrtaceae family and is mostly found in Australia, encompassing approximately 100 species native to the region (Vieira 2004). This family is composed of 130 genera and approximately 4,000 species distributed mainly in areas of tropical and subtropical climate (Cronquist 1981; Souza and Lorenzi 2005), and has been predominantly diversified in Australia and South America (Wilson et al. 2001; Monteiro et al. 2014). In Brazil, the family Myrtaceae represents one of the largest families of plants, with about 23 genera and approximately 1,000 species (Landrum and Kawasaki 1997; Souza and Lorenzi 2005; Gressler et al. 2006).

The species *Melaleuca alternifolia* is native to Australia and it's described as a small tree that grows up to 7m in height, has a paper-like bark and sharp narrow leaves of up to 20mm in length (Carson et al. 2006). This species tolerates diverse environmental conditions and broad range of soils in sub-tropical climates, showing better growth in moist soils and bright environment (Scharifi-Rad et al. 2017).

The essential oils are considered a hallmark in the Myrtaceae family, one of their main characteristics is the presence of oil secretory cavities in their vegetative organs and, in this case, they have taxonomic and systematic anatomy importance (Metcalfe and Chalk 1950; Metcalfe and Chalk 1983). In the genus *Melaleuca*, the secretory system of oils is in the leaves, being composed of numerous cavities located near the epidermis (Fahn 1979; Roshchina and Roshchina 1993). Previous studies using histolocation have confirmed these production sites and indicate that the storage of these essential oils also occurs in these structures (Silva 2007).

For species of aromatic plants, the variations in chemical components of their essential oils are used as a form of identification of different chemotypes. In Melaleuca it is possible to find different chemotypes in distinct populations of many species, so in this case oils that come from plants that are cultivated in Brazil might have a different composition of secondary metabolites as those that were cultivated in (or are native from) other environments, depending on various environmental factors that can exert great influence on the production and chemical composition of essential oils, such as the duration and intensity of stress (be it hydric or caused by injuries to the subject), the stage of development of the plant, nutritional conditions, temperature, air pollution, luminosity, climate, elevation, latitude, type of water system, relative humidity, total duration of time exposed to wind (Southwell and Lowe 2003; Simões et al. 2017) besides other factors which are unrelated to plants' environment such as the entire process of which the oil is obtained, the time of year and the period of the day in which the plant was harvested, as well as the inherent genetic characteristics of the plant, which in the end leads to a difference in the commercial value of production and the quality/quantity of essential oil produced from this plant (Brophy et al. 1989; Southwell and Lowe, 2003; Silva 2007; Simões et al. 2017; Sharifi-Rad et al. 2017).

Modern theories suggest that all secondary metabolites are produced as a result of stimuli, mainly as a defense mechanism against herbivory, and these metabolites are integrated with the adequate receptor systems (Zhao et al. 2005). In *M. alternifolia* something that explains the smaller concentration of oil in the stems is the presence of a suberified epidermis as a barrier against herbivores (since the oil protects against predators, these suberified structures don't need to produce it as much). However, a comparative study of specimens of *Melaleuca* occurring in their natural habitat and locations with dry soil is necessary to establish the correlation of oil structures to environmental or genetic factors. Other studies can benefit from cell culture techniques to verify this influence on the composition of essential oils (Paschen et al. 2006).

The different species of *Melaleuca* present different patterns in relation to their respective essential oil composition. In *M. alternifolia*, the essential oil contains at least 97% of mono and sesquiterpenes, and out of these the main component is terpinen-4-ol (Brophy et al. 1989; Silva 2007), which can consist of up to 53.7% of the total composition of the oil from the leaves. The average yield of the oil extraction from the leaves is of approximately 4% whereas for the stems it's around 0.8% (Silva 2007).

In *M. alternifolia* stems we can find compounds in lower contents, such as: terpinen-4-ol (24.2%), α-terpineol (2.2%), tetradecane (5.3%), β-(Z)-farnesene (10.2%), β-(E)-farnesene (3.6%), viridiflorene (5.4%), δ-cadinene (4.4%), ledol (2.2%), spathulenol (2%), globulol (14.2%) and viridiflorol (7.5%), which amounts to total of 81.2% of this essential oil's composition. While the oil extracted from the leaves, have as minority components: terpinen-4-ol (53.7%), γ-terpinene (18.9%), terpinene (7.6%), p-cymene (3.7%), α-terpineol (3.7%), α-terpinolene (3%), globulol (2%) and viridiflorol (1.6%), which amounts to a total of 94.2% of this oil's composition (Silva 2007). *M. alternifolia* essential oil, in this study, presented high concentrations of terpinen-4-ol and low concentrations of 1,8-cineol.

These numbers where very similar to others found in works of Lin et al. (2016), Wang et al. (2017) and Swords and Hunter (1978), although these values varied slightly, and, the latter work reported the presence of limonene (1%), 1,8-cineol (5.6%), aromadendrene (2.7%), two unknown sesquiterpenes (1.6% each) and viridiflorol (1.03%) (the last one at the time was the first reported case of this substance being found in nature), amongst other substances found in smaller concentrations in the oil.

Relating to the enantiomeric composition of TTO, it was demonstrated that it is possible to distinguish between some types of enantiomers, the results of NMR (nuclear magnetic resonance spectroscopy) and GC (gas chromatography) for terpinen-4-ol (76:24; 72:28) and α-terpineol (35:65; 33:67) confirm the purity of enantiomers. One possible application of this analytic methodology is demonstrated by the comparison between analyses in the quality of different *M. alternifolia* oils (Leach et al. 1993).

Regarding the intraspecific variation of essential oil of *M. alternifolia*, terpenes analysis have revealed differences in the quality of the oil's composition both intrapopulation and interpopulation. There are six naturally occurring chemotypes, that differ in terpene profiles (Homer et al. 2000; Groot and Schmidt 2016; Padovan et al. 2017). Among them, three chemotypes are more common: rich in terpinen-4-ol (the "Type" form), 1,8-cineol or terpinene, whereas the other three types differ distinctly either in the concentrations of terpinen-4-ol or terpinene, but are also rich in 1,8-cineol. There is some evidence that a high intraspecific variation in the composition of terpenes can be beneficial in the effectiveness of chemical defenses against herbivory. This differential resistance against predators can result in a directional selection towards specific terpenes that are toxic to such animals (Butcher et al. 1994).

ANTIBACTERIAL ACTIVITY OF TEA TREE ESSENTIAL OIL

TTO has been widely investigated for its antimicrobial activities, especially against multidrug-resistant bacteria. Carson et al. (1995), recognizing the risks of hospital infections due to the dissemination of Methicillin-resistant *Staphylococcus aureus* (MRSA), as well as the emergence of Mupirocin-resistant MRSA and the subsequent difficulty in the elimination of the dissemination of such pathogens, tested the antibacterial activity of TTO against 66 isolates of *S. aureus*, of which 64 were resistant to methicillin (MRSA) and 33 were resistant to mupirocin. They showed considerable TTO susceptibility at 60 isolates independently of their resistance to mupirocin. These studies prompted more works regarding the antimicrobial activities of TTO and emphasized its importance as a potential agent in the control of antibiotic-resistant bacteria.

Similarly, Nelson (1997) used TTO in susceptibility tests in MRSA and vancomycin-resistant *Enterococcus faecium* (VRE) and indicated the oil's efficacy against isolates tested. May et al. (2000) tested the antimicrobial activity of standard TTO and TTO extracted from a tea tree modified (clone 88) in order to increase its antimicrobial activity against MRSA isolates and Methicillin-sensitive *Staphylococcus aureus* (MSSA) isolates. Their results showed a greater MSSA susceptibility towards both oils and a greater MRSA resistance against the standard oil when compared to the oil extracted from tea tree modified (clone 88).

Brady et al. (2006) compared the susceptibility of MRSA, MSSA and coagulase-negative *Staphylococcus* (CoNS) against tea tree oil and, unlike the results obtained in previous studies (Carson et al. 1995; Nelson 1997; May et al. 2000), the isolates did not present significant sensitivity. However, TTO showed activity against the strains when exposed to high TTO concentrations.

Still regarding resistant microorganisms, Kulkarni et al. (2012) tested TTO activity against uropathogenic *Escherichia coli*, *Klebsiella pneumoniae*, *Proteus mirabilis*, *Proteus vulgaris*, *Pseudomonas aeruginosa* and *S. aureus*, that were resistant to broad-spectrum antibiotics, mainly ampicillin. Their results showed a significant susceptibility of the isolates to TTO, with an emphasis to *E. coli* (MIC 0,03%), *P. mirabilis* (MIC 0,038%) and *P. vulgaris* (MIC 0,04%), while *P. aeruginosa* was more resistant.

Hammer et al. (1996), when discussing the use of TTO as a component in antiseptic products, determined the susceptibility of a range of transient and commensal skin microbiota to TTO. Based on the minimal inhibitory concentrations (MICs) and minimum bactericidal concentrations (MBCs), it was possible to identify a greater susceptibility to TTO from gram-negative bacteria such as *Acinetobacter baumannii*, *Klebsiella pneumoniae* and *S. aureus*, in relation to *Micrococcus luteus*, *M. varians* and *Micrococcus* spp. These results showed the greater efficacy of Melaleuca oil against transient microorganisms when compared to commensal microbiota. This was a great leap in the confirmation of the efficiency of antiseptic soaps containing TTO which could be widely used in hand washing by medical professionals such as surgeons.

Following the example of earlier studies and aiming to prove the efficacy of TTO in oral care products, Hammer et al. (2003) determined the sensitivity of oral bacteria isolates to TTO in concentrations that varied from 4 to 0.004%. This work showed that, when compared to *Streptococcus* spp. isolates, which presented MICs between 0,12-2% and MBCs between 0,5-2%, *Prevotella* spp. and *Veillonella* spp. isolates had greater susceptibility to TTO, with MICs and MBCs of 0.016%. Tests conducted under time-kill model showed that the killing time for *Streptococcus mutans* and *Lactobacillus rhamnosus* isolates was 30 seconds after exposition to TTO at a concentration of 0.5%, although there is no evidence of same results *in vivo*. Other experiments also confirmed the antimicrobial potential of this oil in such products (Groppo et al. 2002; Kamath et al. 2019).

Graziano et al. (2016) comparatively assessed the activities of TTO and chlorhexidine growth of oral bacteria strains *Porphyromonas gingivalis* and *Porphyromonas endodontalis* and of the volatile sulfur compounds (VSCs) produced by them. TTO presented antimicrobial activity against both isolates and also decreased the production of CH_3SH by *P. gingivalis* and especially decreased the production of H_2S and CH_3SH by *P. endodontalis*, which are the main compounds responsible for intra-oral halitosis (Akaji et al., 2014), besides promoting the development of periodontal disease (Nakano et al. 2002; Makino et al. 2012). Chlorhexidine is a compound that is already used in mouthwashes and, as expected, it also presented antibacterial action against the assessed strains, however, this product also showed several side effects such as alteration in taste, teeth coloration and burning sensation on the buccal mucosa. TTO, as a natural product with bactericidal efficacy under low concentrations that are similar to chlorhexidine, offer a great potential in use as a component in oral antiseptics.

Nowadays, TTO has been implemented in several biotechnological development studies. As an example, Ge et al. (2015), through solution casting methods, produced chitosan-based films loaded with TTO droplets

(TTO/CS) and assessed this films' activity against skin pathogens and verified its efficacy against the gram-positive bacteria *Staphylococcus aureus* and the gram-negative bacteria *Escherichia coli*, of which demonstrated activity. They also observed that the films' activity increased expressively with the proportional increase of TTO content present. However, taking into consideration that the chitosan/TTO ratio of 20:4 maintained the hemostatic effect of the pure chitosan film whilst also presenting a broad-spectrum antimicrobial action, the researchers determined this ratio as the ideal. Furthermore, they found that the TTO-CS film was indeed non-toxic and could support fibroblast growth, which is something that puts this film as a suitable material for uses in wound treatments and healing. However, more *in vivo* tests are necessary in order to ensure this product's functionality.

Souza et al. (2014) affirmed that although it is a great antimicrobial agent, TTO presents great limitations due to its physical properties, mainly stemming from its solubility and volatility which can influence the final product in an unstable manner. Because of that, they emphasized the importance of techniques such as nanoencapsulation of TTO as a means to ensure its therapeutic efficacy due to the gradual release of the drug and decrease in any possible toxicity. In order to do that, the researchers assessed the antimicrobial activity of TTO lipidic nanoparticles against pathogenic bacteria and fungi. The susceptibility of the genus *Mycobacterium* against the nanoencapsulated TTO was tested, however, only *M. smegmatis* was shown to be susceptible. In general, when compared to the antimicrobial action provided by pure TTO, higher concentrations of TTO nanoparticles were necessary in order to have an expressive activity. However, despite these results not being what was expected, this study was seen in a positive manner with the confirmation of the maintenance of TTO's antibacterial and antifungal activities, which could influence further research.

It is clear that TTO provides fairly positive results when it comes to its antibacterial activity, which includes activity against broad-spectrum antibiotic-resistant pathogenic bacteria (or resistant to antibiotics that are widely used against certain microorganisms). Nonetheless, TTO's complexity as well as its physical properties may negatively interfere with the formulation of products with this oil since we still have no way to ensure its stability, especially when used *in vivo*, however, studies regarding the antimicrobial activity of TTO still remain very promising.

ANTIFUNGAL ACTIVITY OF TEA TREE ESSENTIAL OIL

The tea tree essential oil, besides presenting broad antibacterial activity, is also fairly efficient against several species of fungi such as: *Candida albicans* (Ge et al. 2015; Pachava et al. 2015; Ramadan et al. 2019), *Aspergillus niger* (An et al. 2019), *Botrytis cinerea* (Yu et al. 2015; Li et al. 2017; Xu et al. 2017b), *Penicillium expansum* (Li et al. 2017; Neto et al. 2019), *Malassezia pachydermatis* (Weseler et al. 2002), *Trichophyton rubrum* (Flores et al. 2013), *Aspergillus ochraceus* (Kong et al. 2019), *Trichophyton equinum* (Pisseri et al. 2009), *Rhizopus stolonifer* (Shao et al. 2013a), amongst others.

Besides from that, several authors have assessed the antifungal activity of TTO, alongside with its main components (terpinen-4-ol, 1,8-cineol, γ-terpinene, α-terpineol, etc.) (Silva 2007). *In vitro* and *in vivo* tests show that the main component of this essential oil, terpinen-4-ol, surpasses the effects of the oil itself, indicating that the oil's inhibitory effect stems mostly from this compound (D'Auria et al. 2001; Mondello et al. 2006; Mertas et al. 2015; Yu et al. 2015; Li et al. 2017; An et al. 2019).

Other papers also assessed TTO's efficacy against fungi that present low susceptibility, such as *Pythium insidiosum* (Valente et al. 2016) or resistance to the widely available antifungals, such as fluconazole-resistant *Candida albicans* (Mertas et al. 2015) and other strains that are resistant to azolic compounds (fluconazole and itraconazole) (Mondello et al. 2006). Mondello et al. (2006) reported the potential of some bioactive compounds in TTO against *C. albicans*. In this case for both compounds used in the study, terpinen-4-ol and eucalyptol (1,8-cineole), there was inhibition of all fungal strains, including those that were resistant to fluconazole and itraconazole.

The application of TTO in dentistry was proposed by Pachava et al. (2015) as a new therapeutic alternative for the stomatitis treatment, due to its activity against the *C. albicans*, the causative agent of this disease. Ge et al. (2015) incorporated TTO to chitosan films as a means to evaluate the efficacy in the treatments against *C. albicans*. Their results show that the incorporation of this EO in antimicrobial films (e.g., chitosan films) was positive and presented a broad action spectrum against the isolates that were tested (bacteria and fungi), which indicates a possible application of these films as wound treatments. More recently, Ramadan et al. (2019) used the TTO as a nanoemulsion combining silver nanoparticles (AgNP) against *C. albicans* and observed its inhibitory activities of the TTO against this species.

In the literature we can find several studies regarding the antifungal activity of TTO towards phytopathogenic fungi, which can cause several problems to crops. *Botrytis cinerea*, the causal agent of the gray mold rot, is one the most destructive pathogens of vegetables and fruits both in the field and during storage (Abbey et al., 2019). Yu et al. (2015) and Li et al. (2017) verified the effect of TTO against *B. cinerea*, and in both cases TTO caused mycelial inhibition (which was concentration-dependant). However, Yu et al. (2015) assessed the inhibitory action of the components from this essential oil (terpinen-4-ol, α-terpineol, 1,8-cineol, terpinolene, etc.) individually and indicated a stronger inhibitory activity of terpinen-4-ol when compared to the oil itself.

A great number of research papers report the use of TTO to minimize problems caused by *B. cinerea* as well as other fungal species in strawberry crops. Shao et al. (2013b) assessed the efficacy of volatilized TTO towards *B. cinerea* and *Rhizopus stolonifer* (black bread mold) and it showed strong antifungal action against both fungi, as well as promoted certain resistance to the strawberries against diseases. Under another perspective, Wei et al. (2018) used hot air treatment alongside TTO against *B. cinerea*. This combination of techniques was shown to be very effective in the treatment against this fungus both *in vitro* and *in vivo* tests, in which they allowed the fruits to maintain their post-harvest quality.

Terzi et al. (2007) verified the *in vitro* action of TTO and its components (terpinen-4-ol, γ-terpinene and 1,8-cineole) against the fungi *Fusarium graminearum*, *Fusarium culmorum* and *Pyrenophora graminea*. They found that TTO and its compounds showed antifungal activities towards all isolates, and terpinen-4-ol was the most effective.

Other fungi such as *Aspergillus ochraceus* and *Aspergillus niger* can cause losses in post-harvest fruits such as grapes. Kong et al. (2019) and An et al. (2019) studied the inhibitory potential of tea tree essential oil and its components against these fungi, and showed that terpene-4-ol showed the best antifungal activity. TTO's inhibitory potential against *Penicillium expansum* was demonstrated by Li et al. (2017), however, Neto et al. (2019) reported recently low susceptibility of this species to the oil *in vitro* tests. The divergences in these results might be due to the low standardization in the concentration of compounds present in the commercially available oils.

There are several works that show the applications of TTO using other products and techniques being already in use for some treatments, such as nanoemulsion, nanoencapsulation and silver nanoparticles. Flores et al. (2013) obtained promising results using nanoencapsulated TTO against the dermatophyte fungus *Trichophyton rubrum*. In their study the use of nanoencapsulated TTO was the most promising result, indicating the strong potential of this release mechanism in the treatment of dermatophytosis caused by this fungus.

Pisseri et al. (2009) used sweet almond oil as an excipient for the application of TTO in the treatment of equines affected by dermatophytosis caused by the fungus *Trichophyton equinum*. The researchers verified that the mix between TTO/sweet almond oil (1:3) was efficient against the fungus.

Other papers also reported the antifungal activity of TTO against other dermatophytic fungi such as *Malassezia pachydermatis* (Weseler et al., 2002) and *Trichophyton mentagrophytes* (Inouye et al. 2006), showing the high potential of the application of TTO as an alternative agent in veterinary medicine. Ramadan et al. (2019) assessed the use of TTO as a nanoemulsion, combining silver nanoparticles (AgNP) and TTO towards a broad-spectrum of microorganisms. One of the antifungal activities identified was against *T. mentagrophytes*, where both products tested (TTO and TTO+AgNP) were able to inhibit the fungus and the oil on its own showed the best efficacy.

Valente et al. (2016) also studied the use of the oil and nanoemulsions prepared using TTO against the *Pythium insidiosum* fungus, which presents a low susceptibility against available antifungals. Both demonstrated an inhibitory effect *in vitro* tests conducted, but the nanoemulsion was the one that was shown to be the most effective, which can be justified by the instability/volatility of the components present in the EO. These data reinforce the use of nanoparticles and nanoemulsions as efficient tools in the application of TTO in treatments of several fungal and bacterial infections.

The tea tree essential oil was studied for its antifungal activities against other genera and species such as *Microsporum* sp., *Schizosaccharomyces pombe* and *Debaryomyces hansenii* (D'Auria et al. 2001); *Fusarium oxysporum* (Souza et al. 2014); *Aspergillus fumigatus*, *Cladosporium* sp., *Epidermophyton floccosum* (Hammer et al. 2002), among others.

CONCLUSION

Based on the different stages analyzed, it is possible to confirm that the essential oil of TTO (*Melaleuca alternifolia*) has a high antimicrobial activity, with its major constituents being responsible. Several products with antimicrobial activities of TTO have been demonstrated, however, the major limitation of its application is its volatilization, being necessary the development of formulations that break this unfeasibility. Although TTO has high antimicrobial activity *in vitro*, studies to verify its efficacy *in vivo* should be performed, as its possible routes of metabolization.

REFERENCES

Abbey, J. A., Percival, David., Abbey, L., Samuel, K. Asiedu., Prithiviraj, B. & Schilder, A. 2019. "Biofungicides as alternative to synthetic fungicide control of grey mould (*Botrytis cinerea*) – prospects and challenges". *Biocontrol Science and Technology.*, *29*(3), 207-228. Doi:10.1080/09583157.2018.1548574.

Akaji, E. A., Folaranmi, N. & Shiwaju, O. (2014). "Halitosis: a review of the literature on its prevalence, impact and control". *Oral Health & Preventive Dentistry.*, *12*(4), 297–304. Doi: 10.3290/j.ohpd.a33135.

An, P. X., Yang, J. Yu., Qi, J., Ren, X. & Kong, Q. (2018). "α-terpineol and terpene-4-ol, the critical components of tea tree oil, exert antifungal activities *in vitro* and *in vivo* against *Aspergillus niger* in grapes by inducing morphous damage and metabolic changes of fungus." *Food Control.*, *98*, 42-53. Doi:10.1016/ j.foodcont.2018.11.013.

Bajpai, V. K., Baek, K. H. & Kang, S. C. (2012). "Control of *Salmonella* in foods by using essential oils: a review". *Food Research International.*, *45*(2), 722-734. Doi: 10.1016/j.foodres.2011.04.052.

Bhargava, V. V., Patel, S. C. & Desai, K. S. (2013). "Importance of terpenoids and essential oils in chemotaxonomic approach". *International Journal of Herbal Medicine*, *1*(2), 14-21. Accessed June 30, 2019. http://www.florajournal.com/vol1issue2/4.1.html.

Brady, A., Loughlin, R., Gilpin, D., Kearney, P. & Tunney, M. (2006). "*In vitro* activity of tea-tree oil clinical skin isolates of meticillin-resistant and sensitive *Staphylococcus aureus* and coagulase-negative staphylococci growing planktonically and as biofilms". *Journal of Medical Microbiology.*, *55*, 1375–80. Doi: 10.1099/jmm.0.46558-0.

Brophy, J. J., Davies, N. W., Southwell, I. A., Stiff, I. A. & Lyall R. W. (1989). "Gas chromatographic quality control for oil of *Melaleuca* terpinen-4-ol type (Australian tea tree)". *J. Agric. Food. Chem.*, *37*, 1330-1335. Doi: 10.1021/jf00089a027.

Butcher, P. A., Doran, J. C. & Slee, M. U. (1994). "Intraspecific variation in leaf oils of *melaleuca alternifolia* (Myrtaceae)". *Biochemical*

Systematics and Ecology., 22(4), 419-430. Doi: 10.1016/0305-1978(94)90033-7

Carson C. F., Cookson, B. D., Farrelly, H. D. & Riley, T. V. (1995). "Susceptibility of methicillin-resistant *Staphylococcus aureus* to the essential oil of melaleuca alternifolia brief reports susceptibility of methicillin-resistant *staphylococcus aureus* to the essential oil of *Melaleuca alternifolia*". *Journal of Antimicrobial Chemotherapy*, 35(3), 421–424. Doi: 10.1093/jac/35.3.421.

Carson, C. F., Hammer, K. A. & Riley, T. V. (2006). "*Melaleuca alternifolia* (Tea Tree) oil: a review of antimicrobial and other medicinal properties". *Clinical Microbiology Reviews.*, 19, (1), 50-62. Doi: 10.1128/CMR.19.1.50-62.

Cronquist, A. (1981). *An Integrated System of Classification of Flowering Plants*, Columbia: New York.

D'auria, F. D., Laino, L., Strippoli, V., Tecca, M., Salvatore, G., Battinelli, L. & Mazzanti, G. (2001). "*In vitro* activity of tea tree oil against *candida albicans* mycelial conversion and other pathogenic fungi". *Journal of Chemotherapy.*, 13(4), 377-383. Doi: 10.1179/joc.200 1.13.4.377.

Ehlert, P. A. D., Ming, L. C., Marques, M. O. M., Fenandes, D. M., Rocha, W. A., Luz, J. M. Q. & Silva, R. F. (2013). "Influência do horário de colheita sobre o rendimento e composição do óleo essencial de erva-cidreira brasileira (*Lippia alba* (Mill.) N. E. Br.]". *Revista Brasileira de Plantas Medicinais.*, 15(1), 72-77. Doi: 10.1590/S1516-057220 13000100010.

Fahn, A. (1979). "*Secretory tissues in plants*". London: Academic Press https://catalogue.nla.gov.au/Record/1096427.

Flores, F. C., Lima, J. A., Ribeiro, R. F., Alves, S. H., Rolim, C. M. B., Beck, R. C. R. & Silva, C. B. (2013). "Antifungal activity of nanocapsule suspensions containing tea tree oil on the growth of *Trichophyton rubrum*". *Mycopathologia.*, 175, 281-286. Doi: 10.100 7/s11046-013-9622-7.

Ge, Y. & Ge, M. (2015). "Sustained broad-spectrum antimicrobial and haemostatic chitosan-based film with immerged tea tree oil droplets".

Fibers and Polymers., 16(2), 308–318. Doi: 10.1007/s12221-015-0308-2

Graziano, T. S., Calil, C. M., Sartoratto, A., Franco, G. C. N., Groppo, F. C. & Cogo-Müller, K. C. (2016). "In vitro effects of Melaleuca alternifolia essential oil on growth and production of volatile sulphur compounds by oral bacteria". Journal of Applied Oral Science, 24(6), 582–589. Doi: 10.1590/1678-775720160044.

Gressler, E., Pizo, M. A. & Morellato, L. P. C. (2006). Polinização e dispersão de sementes em Myrtaceae do Brasil. Brazilian Journal of Botany., 29 (4), 509-530. Doi: 10.1590/S0100-84042006000400002.

Groot, A. C. & Schmidt, E. (2016). "Tea tree oil: contact allergy and chemical composition". Contact Dermatitis., 75 (3), 129-143. Doi: 10.1111/cod.12591.

Groppo, F. C., Ramacciato, J. C., Simões, R. P., Flório, F. M. & Sartoratto, A. (2002). "Antimicrobial activity of garlic, tea tree oil, and chlorhexidine against oral microorganisms". International Dental Journal., 52(6), 433–437. Doi: 10.1111/j.1875-595X.2002.tb00638.x.

Hammer, K. A., Carson, C. F. & Riley, T. V. (2002). "In vitro activity of Melaleuca alternifolia (tea tree) oil against dermatophytes and other filamentous fungi". Journal of Antimicrobial Chemotherapy., 50(2), 195–199. Doi: 10.1093/jac/dkf112

Hammer, K. A., Carson, C. F. & Riley, T. V. (1996). "Susceptibility of transient and commensal skin flora to the essential oil of Melaleuca alternifolia (Tea Tree Oil)". American Journal of Infection Control., 24(3), 186–189. Doi: 10.1016/S0196-6553(96)90011-5.

Hammer, K. A., Dry, L., Johnson, M.., Michalak, E. M., Carson, C. F. & Riley, T. V. (2003). "Susceptibility of oral bacteria to Melaleuca alternifolia (Tea Tree) Oil In Vitro". Oral Microbiology and Immunology., 18(6), 389–392. Doi: 10.1046/j.0902-0055.2003.00105.x

Homer, L. E., Leach, D. N., Lea, D., Lee, L. S., Henry, R. J. & Baverstock, P. R. (2000). "Natural variation in the essential oil contents of Melaleuca alternifolia Cheel (Myrtaceae)". Biochemical Systematics and Ecology., 28 (4), 367-382. Doi: 10.1016/S0305-1978(99)00071-x.

Inouye, S., Nishiyama, Y., Uchida, K., Hasumi, Y., Yamaguchi, H. & Abe, S. (2006). "The vapor activity of oregano, perilla, tea tree, lavender, clove, and geranium oils against a *Trichophyton mentagrophytes* in a closed box". *Journal of Infection and Chemotherapy.*, *12*(6), 349-54. Doi: 10.1007/s10156-006-0474-7

Kamath, N. P., Tandon, S., Nayak, R., Naidu, S., Anand, P. S. & Kamath, Y. S. (2019). "The Effect of *Aloe Vera* and Tea Tree Oil Mouthwashes on the Oral Health of School Children". *European Archives of Paediatric Dentistry.*, 1–6. Doi: 10.1007/s40368-019-00445-5.

Kong, Q., Zhang L., An, P., Qi, J., Yu, X., Lu, J. & Ren, X. (2019). "Antifungal mechanisms of a-terpineol and terpene-4-alcohol as the critical components of *Melaleuca alternifolia* oil in the inhibition of rot disease caused by *Aspergillus ochraceus* in postharvest grapes". *Journal of Applied Microbiology.*, *126*(4), 1161-1174. Doi: 10.1111/jam.14193.

Kulkarni, A., Jan, N. & Nimbarte, S. (2012). "Monitoring of antimicrobial effect of GC-MS standardized *Melaleuca alternifolia* oil (tea tree oil) on multidrug resistant uropathogens". *IOSR Journal of Pharmacy and Biological Sciences*, *2*(2), 1-9. Doi: 10.9790/3008-0220614.

Landrum, L. R. & Kawasaki, M. L. (1997). "The genera of Myrtaceae in Brazil - an illustrated synoptic treatment and identification keys". *Brittonia*, *49* (4), 508-536. Doi: 10.2307/2807742.

Leach, D. N., Wyllie, S. G., Hall, J. G. & Kyratzis, I. (1993). "Enantiomeric composition of the principal components of the oil of *Melaleuca alternifolia*". *J Agric. Food Chem.*, *41*(10), 1627-1632. Doi: 10.1021/jf00034a020.

Li, W. R., Li, H. L., Shi, Q. S., Sun, T. L., Xie, X. B., Song, B. & Huang, X. M. (2016). "The dynamics and mechanism of the antimicrobial activity of tea tree oil against bacteria and fungi". *Applied Microbiology and Biotechnology.*, *100*(20), 8865–8875. Doi: 10.1007/s00253-016-7692-4.

Li, Y., Shao, X., Xu, J., Wei, Y., Xu, F. & Wang, H. (2017b). "Effects and possible mechanism of tea tree oil against *Botrytis cinerea* and

Penicillium expansum in vitro and in vivo test". *Canadian Journal of Microbiology.*, *63*(3), 219-227. Doi: 10.1139/cjm-2016-0553.

Lin, P. C., Lee, J. J. & Chang, I. Jy. Chang. (2016). "Essential oils from Taiwan: Chemical composition and antibacterial activity against *Escherichia coli*". *Journal of Food and Drug Analysis.*, *24*(3), 464-470. Doi: 10.1016/j.jfda.2015.12.006.

Makino, Y., Yamaga, T., Yoshihara, A., Nohno, K. & Miyazaki, H. (2012). "Association Between Volatile Sulfur Compounds and Periodontal Disease Progression in Elderly Non-Smokers". *Journal of Periodontology.*, *83*, 635–43. Doi: 10.1902/jop.2011.110275.

May, J., Chan, C. H., King, A., Williams, L. & French, G. L. (2000). "Time-Kill studies of tea tree oils on clinical isolates". *Journal of Antimicrobial Chemotherapy*, *45*(5), 639–643. Doi: 10.1093/jac/45.5.639

Mertas, A., Garbusińska, A., Szliszka, E., Jureczko, A., Kowalska, M. & Król, W. (2015). "The influence of tea tree oil (*Melaleuca alternifolia*) on fluconazole activity against fluconazole-resistant *Candida albicans* Strains". *BioMed Research International*, 1-9. Doi: 10.1155/ 2015/ 590470.

Metcalfe, C. R. & Chalk, L. (1950). *Anatomy of the Dicotyledons: Leaves, stem, and wood in relation to taxonomy with notes on economic uses.* Clarendon: Oxford. https://archive.org/details/anatomyofthedico 03355 2mbp/page/n5.

Metcalfe, C. R. & Chalk, L. (1983). *Anatomy of the Dicotyledons: Wood Structure and Conclusion of the General Introduction.* New York: Oxford University Press. https://link.springer.com/article/ 10.2307/ 2806611.

Mondello, F., Bernardis, F., Girolamo, A., Cassone, A. & Salvatore, G. (2006). "*In vivo* activity of terpinen-4-ol, the main bioactive component of *Melaleuca alternifolia* Cheel (tea tree) oil againstazole-susceptible and resistant human pathogenic *Candida* species". *BMC Infectious Diseases*, *6*(158), 1-8. Doi: 10.1186/1471-2334-6-158.

Monteiro, M. H. D., Macedo, H. W., Silva Junior, A. & Paumgartten, F. J. (2014). "Óleos essenciais terapêuticos obtidos de espécies de

Melaleuca L. (*Myrtaceae* Juss.)".*Revista Fitos.*, *8*(1), 19-32. http://revistafitos.far.fiocruz.br/index.php/revista-fitos/article/view/191.

Nakano, Y., Yoshimura, M. & Koga, T. (2002). "Correlation oral malodor and periondontal bactéria". *Microbes Infect.*, *4*(6), 679-683. Doi: 10.1016/S1286-4579(02)01586-1.

Nelson, R. (1997). "*In-vitro* activities of five plant essential oils against methicillin- resistant *Staphylococcus aureus* and Vancomycin-Resistant *Enterococcus faecium*". *Journal of Antimicrobial Chemotherapy.*, *40*(2), 305–306. Doi: 10.1093/jac/40.2.305.

Neto, A. C. R., Bruno B. Navarro., Canton, L., Maraschin, M. & Di Piero, R. M. (2019). "Antifungal activity of palmarosa (*Cymbopogon martinii*), tea tree (*Melaleuca alternifolia*) and star anise (*Illicium verum*) essential oils against *Penicillium expansum* and their mechanisms of action". *LWT Food Science and Technology.*, *105*, 385-392. Doi: 10.1016/j.lwt.2019.02.060.

Pachava, K. R., Nadendla, L. K., Alluri, L. S. C., Tahseen, H. & Sajja, N. P. (2015). "*In vitro* antifungal evaluation of denture soft liner incorporated with tea tree oil- a new therapeutic approach towards denture stomatitis". *Journal of Clinical & Diagnostic Research.*, *9*(6), 62-64. Doi: 10.7860/JCDR/2015/12396.

Padovan, A., Keszei, A., Yasmin, H., Krause, S. T., Köllner, T. G., Degenhardt, J., Gershenzon, J., Külheim, C., Foley, W. J. 2017. "Four terpene synthases contribute to the generation of chemotypes in tea tree (*Melaleuca alternifolia*)". *BMC Plant Biology.*, 17, 160. Doi: 10.1186/s12870-017-1107-2

Paschen, R., Wells, D., Blair, M. & Sirvent, T. M. (2006). "Optimization of tissue culture conditions of *Melaleuca alternifolia* for enhanced oil production". *Abstracts of Papers, 231st ACS National Meeting*, Atlanta, GA, United States, March 26-March. https://www.acsmedchem.org/ama/orig/abstracts/mediabstracts2006.pdf.

Pisseri, F., Bertoli, A., Nardoni, S., Pinto, L., Pistelli, L., Guidi, G. & Mancianti, F. (2009). "Antifungal activity of tea tree oil from *Melaleuca alternifolia* against *Trichophyton equinum*: An *in vivo* assay." *Phytomedicine.*, *16*(11), 1056. Doi: 1058.10.1016/ j.phymed.2009.03.013.

Ramadan, M. A., Shawkey, A. E., Rabeh, M. A. & Abdellatif, A. O. (2019). "Promising antimicrobial activities of oil and silver nanoparticles obtained from *Melaleuca alternifolia* leaves against selected skin-infecting pathogens". *Journal of Herbal Medicine.*, *16*, 1-20. Doi: 10.1016/j.hermed.2019.100289.

Roshchina, V. V. & Roshchina, V. D. (1993). "The excretory function of higher plants". Springer Verlag. https://link.springer.com/book/10.1007%2F978-3-642-78130-8.

Sanli, A. & Karadoğan, T. (2016). "Geographical Impact on Essential Oil Composition of Endemic *Kundmannia anatolica*". *Afr. J. Tradit. Complement. Altern Med.*, *14*(1), 131–137. Doi: 10.21010/ ajtcam.v14i1.14

Shao, X., Cheng, S., Wang, H., Yu, D. & Mungai, C. (2013a). "The possible mechanism of antifungal action of tea tree oil on *Botrytis cinerea*". *Journal of Applied Microbiology.*, *114*(6), 1642-1649. Doi: 10.1111/jam.12193.

Shao, X., Wang, H., Xu, F. & Cheng, S. (2013b). "Effects and possible mechanisms of tea tree oil vapor treatment on the main disease in postharvest strawberry fruit". *Postharvest Biology and Technology*, *77*, 94-101. Doi: 10.1016/j.postharvbio.2012.11.010.

Sharifi-Rad, J., Salehi, B., Varoni, E. M., Sharopov, F., Yousaf, Z., Ayatollahi, S. A., Kobarfard, F., Sharifi-Rad, M., Afdjei, M. H., Sharifi-Rad, M. & Iriti, M. (2017). "Plants of the melaleuca genus as antimicrobial agents: from farm to pharmacy". *Phytotherapy Research.*, *31*(10), 1475–1494. Doi:10.1002/ptr.5880.

Silva, C. J. (2007). *Morfoanatomia foliar e composição química dos óleos essenciais de sete espécies de Melaleuca L. (Myrtaceae). Master's thesis.* Universidade Federal de Viçosa. http://www.locus.ufv.br/bitstream/handle/123456789/2567/texto%20completo.pdf?sequence=1&isAllowed=y.

Simões, C. M. O., Schenkel, E. P., Mello, J. P. C. & Mentz, L. (2017). *Farmacognosia: do produto natural ao medicamento.*, *172*. Porto Alegre: Artmed.

Southwell, I. & Lowe, R. (2003). "*Tea Tree: The Genus Melaleuca*". CRC Press: Boca Raton, Florida, United States. 63. https://www.crc press. com/Tea-Tree-The-Genus-Melaleuca/Southwell-Lowe/p/book/978905 7024177.

Souza, V. C. & Lorenzi, H. (2005). Botânica Sistemática: *Guia ilustrado para identificação das famílias de Angiospermas da flora Brasileira, baseado em APG/II.* Nova Odessa, SP: Instituto Plantarum.

Souza, E. M., Lopes, L. Q. S., Vaucher, R. A., Mário, D. N., Alves, S. H., Agertt, V. A., Bianchini, B. V., Felicidade, S. I., Campus, M. M. A., Boligon, A. A., Athayde, M. L., Santos, C. G., Raffin, R. P., Gomes, P. & Santos, R. C. V. (2014). "Antimycobacterial and antifungal activities of *Melaleuca alternifolia* oil nanoparticles". *Journal of Drug Delivery Science and Technology.*, *24*(5), 559-560. Doi:10.1016/ S177 3-2247(14)50105-0.

Swords, G. & Hunter, G. L. K. (1978). "Composition of australian tea tree oil (*Melaleuca alternifolia*)". *J. Agric. Food Chem.*, *26*(3), 734-737. Doi: 10.1021/jf60217a031.

Terzi, V., Morcia, C., Faccioli, P., Vale, G., Tacconi, G. & Malnati, M. (2007). "*In vitro* antifungal activity of the tea tree (*Melaleuca alternifolia*) essential oil and its major components against plant pathogens." *Letters in Applied Microbiology.*, *44*(6), 613-8. Doi: 10.1111/j.1472-765X.2007.02128. x.

Thosar, N., Basak, S. Basak., Bahadure, R. & Rajurkar, M. (2013). "Antimicrobial efficacy of five essential oils against oral pathogens: an in vitro study". *European Journal of Dentistry.*, *7*(5), 71-77. Doi: 10.4103/1305-7456.119078

Valente, J. S. S., Fonseca, A. O. S., Brasil, C. L., Sagave, L., Flores, F. C., Silva, C. B., Sangioni, L. A., Pötter, L., Santurio, J. M., Botton, S. A. & Pereira, D. I. B. (2016). "*In vitro* activity of *Melaleuca alternifolia* (tea tree) in its free oil and nanoemulsion formulations against *Pythium insidiosum*". *Mycopathologia*, (181)11-12, 865-869. Doi: 10.1007/s11046-016-0051-2.

Vieira, T. R., Barbosa, L. C., Maltha, C. R., Paula, V. F. & Nascimento, E. A. (2004). "Constituintes químicos de *Melaleuca alternifolia (Myrtaceae)*." *Química Nova.*, 27(4), 536-539. Doi: 10.1590/S0100-40422004000400004

Wang, H. F., Yih, K. H., Yang, C. H. & Huang, K. F. (2017). "Antioxidant activity and major chemical component analyses of twenty-six commercially available essential oils". *Journal of Food and Drug Analysis*, 25(4), 881-889. Doi: 10.1016/j.jfda.2017.05.007.

Wei, Y., Wei, Y., Xu, F. & Shao, X. (2018). "The combined effects of tea tree oil and hot air treatment on the quality and sensory characteristics and decay of strawberry". *Postharvest Biology and Technology.*, (136), 139-144. Doi: 10.1016/j.postharvbio.2017.11.008.

Weseler, A., Geiss, H. K., Saller, R. & Reichling, J. (2002). "Antifungal effect of Australian tea tree oil on *Malassezia pachydermatis* isolated from canines suffering from cutaneous skin disease". *Schweiz. Arch. Tierheilkd.*, 144(5), 215-221. Doi: 10.1024/0036-7281.144.5.215.

Wilson, P. G., O'Brien M. M., Gadek, P. A. & Quinn, C. J. (2001). "Myrtaceae revisited: a reassessment of infrafamilial groups". *American Journal of Botany.*, 88(11), 2013–2025. Doi:10.2307/355 84 28.

Xu, J., Shao, K., Li, Y., Wei, Y., Xu, F. & Wang, H. (2017^a). "Metabolomic analysis and mode of action of metabolites of tea tree oil involved in the suppression of *Botrytis cinerea*". *Frontiers in Microbiology.*, (8), 1017. Doi: 10.3389/fmicb.2017.01017.

Xu, J., Shao, X., Wei, Y., Xu, F. & Wang, H. (2017b). "iTRAQ Proteomic Analysis reveals that metabolic pathways involving energy metabolism are affected by tea tree oil in *Botrytis cinerea*". *Frontiers in Microbiology.*, (8), 1989. Doi: 10.3389/fmicb.2017.01989.

Yu, D., Wang, J., Shao, X., Xu, F. & Wang, H. (2015). "Antifungal modes of action of tea tree oil and its two characteristic components against *Botrytis cinerea*". *Journal of Applied Microbiology.*, 119(5), 1253-126 2. Doi: 10.1111/jam.12939.

Zhang, X., Guo, Y., Guo, L., Jiang, H. & Ji, Q. (2018). "*In vitro* evaluation of antioxidant and antimicrobial activities of *Melaleuca alternifolia* essential oil". *BioMed Research International.* 1-8. Doi: 10.1155/2018 /2396109.

Zhao, J., Davis, L. C. & Verpoorte, R. (2005). "Elicitor signal transduction leading to production of plant secondary metabolites". *Biotechnology Advances.*, *23*(4), 283-333. Doi: 10.1016/j.biotechadv.2005.01.003.

In: Antimicrobial Potential of Essential Oils   ISBN: 978-1-53616-945-4
Editors: B. Oliveira de Veras et al.   © 2020 Nova Science Publishers, Inc.

Chapter 5

ANTIMICROBIAL ACTIVITY OF ESSENTIAL OILS FROM CAATINGA PLANT SPECIES

José Rafael da Silva Araujo[1,*],
Paulo Henrique Valença Nunes[2],
Camila Marinho da Silva[1], *Suyane de Deus e Melo*[1],
Marx Oliveira de Lima[1], *Silvany de Sousa Araujo*[1]
and Bruno Oliveira de Veras[3]

[1]Department of Genetics, Federal University of Pernambuco, Recife, Pernambuco, Brazil
[2]Department of Pharmacy, Federal University of Pernambuco, Recife, Pernambuco, Brazil
[3]Department of Medicine Tropical, Federal University of Pernambuco, Recife, Pernambuco, Brazil

ABSTRACT

Caatinga is a unique Brazilian cultural formation and much of the biological heritage is not found anywhere in the world. This biome possesses a variety of plant species with the primary vegetation still

---

[*] Corresponding Author's Email: jrafaelquadros@hotmail.com.

preserved in some areas. Its species have physiological characteristics that reflect complex and peculiar adaptations to the exceptional conditions of this environment, which aroused the interest of the scientific community. This study focused on plant species that present essential oils (EOs) with antimicrobial potential and that occurs in the Caatinga Phytogeographical Domain (CPD). Three botanical families with occurrence in CPD were select for the current study: Euphorbiaceae, Asteraceae, and Lamiaceae. β-Caryophyllene was one of the most recurrent compounds, being responsible for this activity. Among the classes of microorganisms analyzed, all families showed activity against Gram-positive bacteria and some species of fungi. However, the Lamiaceae family, besides being effective against the mentioned classes, showed activity against Gram-negative bacteria as well. The expansion of biochemical prospection of Caatinga plants is extremely important since its plant species have been the source of biomolecules that can become new alternatives for antimicrobial drugs.

INTRODUCTION

Caatinga is one of the largest and distinct plant formations in the world, inhabiting more than 25 million people, occupying an area of 800,000 km^2, including the states of Piauí, Ceará, Rio Grande do Norte, Paraíba, Pernambuco, Alagoas, Sergipe, Bahia, and Minas Gerais. It is a unique Brazilian cultural formation and much of the biological heritage is not found anywhere in the world (Leal et al., 2005). Until now, the Caatinga possesses about 4,482 plant species distributed in 1,133 genera and 159 families, with several number of remaining plants still well preserved, including an expressive number of rare and endemic taxa (Forzza et al. 2013). These plant species have physiological characteristics that reflect complex and peculiar adaptations to the unique conditions of this environment, which aroused the interest of the scientific community (Moura et al. 2015).

Part of these plant species, derived from this Brazilian vegetable formation, are used for different medicinal purposes (Albuquerque et al. 2012). These plants are a source of biomolecules that can become new alternatives for the industry: analgesics, tranquilizers, diuretics, laxatives, and antibiotics, among others (Arcoverde et al. 2014). In this sense, there is a stimulation for expansion of the biological and biochemical prospection of Caatinga plants.

Among the secondary metabolites derived from plants, the essential oils (EOs) comprise the group of the aromatic and volatile oil obtained from different plant parts, including flowers, buds, roots, bark, and leaves. Most EOs are composed of terpene compounds (mono-, di-, tri-, and sesquiterpenes), alcohols, acids, esters, epoxides, aldehydes, ketones, amines and sulfides (Dima and Dima 2015; Swamy et al. 2016). The main groups of EOs are divided in two groups: (i) aromatic and aliphatic compounds, and (ii) hydrocarbon terpenes (isoprenes) and terpenoids (isoprenoids) (Bakkali et al. 2008; Eslahi et al. 2018). These compounds can be extracted by expression, fermentation, enfleurage, extraction or distillation (Roohinejad et al. 2018).

EOs exhibits a wide spectrum of biological activities, which includes insecticidal, antioxidant, anti-inflammatory, anti-allergic, and anticancer agents (Raut et al. 2014). Furthermore, many EOs exert strong antibacterial and antifungal activities (Swame et al. 2016; Shojaee-Aliabadi et al. 2018), stimulating their application for alternative strategies for drug-resistant pathogens (Savoia et al. 2012; Langeveld et al. 2014) and also for antimicrobial control in food industry (Jayasena et al. 2013; Calo et al. 2014).

Essential oils, derived from native Caatinga species have presented satisfactory antimicrobial and antifungal activity. Fontenelle et al. (2008) demonstrated that OEs of the different species of *Croton* genus (Euphorbiaceae family) exhibit good antifungal activity against *Microsporum canis* and *Candida* spp. with low toxicity in Wistar mice. Yet, species from Caatinga domain have antibacterial activity against *Aeromonas hydrophila*, *Escherichia coli*, *Listeria monocytogenes*, *Salmonella enteritidis* and *Staphylococcus aureus* and many others (Costa et al. 2013; Araújo et al. 2017).

In this sense, this chapter aimed to gather species that occur in the Caatinga (phytogeographical domain), from the botanical families Euphorbiaceae, Asteraceae, and Lamiaceae, due to their abundance of species in each family and considerable antimicrobial activity. The major classes of the main secondary metabolites have also been listed, which are possibly responsible for this biological activity, important for the development of new drugs.

METHODS

Study of Area

This study focused in the plant species that present essential oils with antimicrobial potential that occur in the Caatinga Phytogeographical Domain (CPD). It is the main phytogeographic domain in the Northeast region of Brazil, composed of different floristic groups and physiognomies. The main vegetation kind of the CPD is the Caatinga *stricto sensu*. The CPD occupies an area of 800,000 km^2 (according to IBGE 2004 has 844,453 km^2) in the states of Bahia, Sergipe, Alagoas, Pernambuco, Paraíba, Rio Grande do Norte, Ceará, Piauí, small area of Maranhão and the north of Minas Gerais (Figure 1). This makes CPD one of the largest semi-arid regions in the world and due to your peculiarities, it is recognized as a unique region in the world (Ab´Sáber 1977; Olson et al. 2001).

Database Compilation

The "List of Species of Brazilian Flora" available at <http://reflora.jbrj.gov.br/reflora/herbarioVirtual > was used to access the CPD plant species. Only species registered in this platform were contemplated. For this, the filters presented in Table 1 were applied to collect the data.

Cultivated, native, naturalized, endemic, and not endemic CPD species were included. Only Euphorbiaceae, Asteraceae, and Lamiaceae family species were analyzed, due to their predominant occurrences in the CPD (Flora do Brasil 2019).

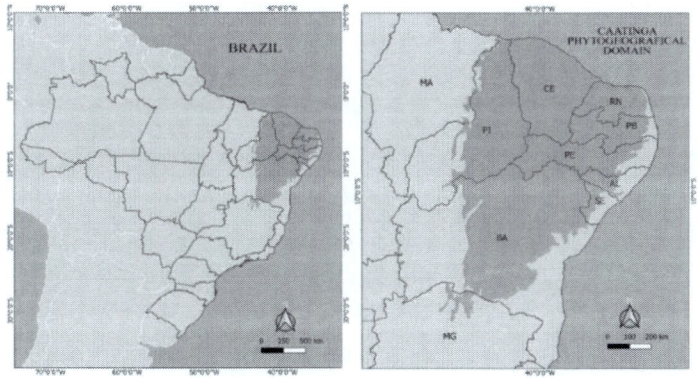

Figure 1. Caatinga Phytogeographical Domain (CPD) in Brazil's northeast, focus of the studies examined in this work (NE Brazil).

Table 1. Parameters considered for the survey of Caatinga Phytogeographical Domain (CPD) plant species through the "List of Species of the Brazilian Flora" platform

	Parameter	Choice
Name	Group	"Angiosperms"
	Family	"Euphorbiaceae", "Asteraceae" or "Lamiacae"
Lifestyle and substrate	Lifestyle	"All"
	Substrate	"All"
Geographic distribution	Occurs in Brazil	"Yes"
	Occurrence	"Occur in"
	Endemism[a]	"All"
	Origin[b]	"All"
Distribution	Region	"Any"
	Restrict to oceanic islands	Not selected
	State	"Any"
	Phytogeographical domains	"Caatinga"
Vegetation		"All"

[a] Endemic and non-endemic CPD plant species were included
[b] Cultivated, native and naturalized Brazil's plant species were included

After the application of the filters, the antimicrobial activity of the essential oils from each plant species was investigated. Google Scholar <https://scholar.google.com.br> and PubMed <https:// www.ncbi.nlm.nih.gov/pubmed> databases were used to found this potential. The keywords "Species name" "Essential Oil" "Antimicrobial" were employed. Only the peer-reviewed studies were considered.

RESULTS AND DISCUSSION

Essential oils have been proved to be a promising source for bioactive molecules with the potential to be an alternative to conventional antibiotics, as well as acting against phytopathogens (Fonseca et al. 2015). In this review, the study of the essential oils of the species that occur in CPD and their respective antimicrobial activities reveals the rich potential of these species in this domain, besides providing the current dimension of the number of studies carried out. To realize this, species from Flora of Brazil Platform where considered, which are cataloged through monographs and represent almost all species that occur in Brazil (Flora do Brasil 2019).

Only the families Asteraceae, Euphorbiaceae, and Lamiaceae were selected for this study, due to the abundance of the species of these families in the phytogeographic domain under study (Flora do Brasil 2019). According to the Flora of Brazil platform, the Euphorbiaceae family presents 229 species distributed in 31 genera with occurrence in the Caatinga. Of this total, 11 species were studied, distributed in four genera. The genus *Croton* presented eight species, followed by *Jatropha* with two species and the genera *Euphorbia* and *Ricinus* with only one species each (Table 2).

For the Asteraceae family, the platform presented 303 species distributed in 118 genera with occurrence in the Caatinga. In this total, 20 species were studied, distributed in 16 genera. The genera *Conyza*, *Ageratum*, *Chromolaena*, and *Pectis* presented two species, while the genera *Tanacetum*, *Baccharis*, *Egletes*, *Solidago*, *Bidens*, *Austroeupatorium*, *Achyrocline*, *Acmella*, *Eclipta*, *Acanthospermum*, *Tridax*, and *Vanillosmopsis* presented one species each (Table 3).

For the Lamiaceae family, the Flora of Brazil platform reports 97 species distributed in 26 genera with occurrence in the Caatinga. In total, 14 species were studied, distributed in eight genera. The genus *Hyptis* presented five species, followed by *Ocimum* with three species, *Leonotis* and *Vitex* with two species and *Hypenia*, *Leonurus*, *Mentha* and *Nepeta* presented one species each (Table 4).

As observed in the results, it is possible to verify that the study of EOs and their antimicrobial potential is scarce for botanical families that occur in CPD: 5% for the Euphorbiaceae, 8% for Asteraceae, and 14% for Lamiaceae. Concerning the diversity of genera studied among each family, the genus *Croton* (Euphorbiaceae family) presents an abundance of studied species. Other genera exhibited few studies species and once they showed great antimicrobial potential, it is recommended that further studies involving these genera species. Although it showed few studies, it is possible to observe the strong antimicrobial potential of EOs from species that occur in this region.

Regarding the antimicrobial survey, the Euphorbiaceae family were tested against 28 microbial species (18 bacteria and 10 fungi). The OEs from these family species showed great effectiveness against Gram-positive bacteria (*Enterococcus faecalis* and *Listeria monocytogenes*) and only one fungus (*Microsporum canis*).

The Asteraceae family species were tested against 67 species (37 bacteria and 30 fungi). The family showed great effectiveness against fungi (*Aspergillus fumigatus*, *Aspergillus niger*, *Candida albicans*, *Penicillium chrysogenum*) and Gram-positive bacteria (*Bacillus cereus*, *Mycobacterium tuberculosis*, and *Staphylococcus aureus*).

Already the Lamiaceae family species, the OEs were tested against 72 species (39 bacteria and 33 fungi). The family showed great effectiveness against Gram-positive bacteria (*Staphylococcus aureus*, *Staphylococcus epidermidis*, *Bacillus cereus*, *Bacillus pumillus*, *Bacillus subtilis*, and *Enterococcus faecalis*), Gram-negative bacteria (*Escherichia coli*, *Salmonella typhimurium*, *Neisseria gonorrhoeae*, *Pseudomonas aeruginosa*, *Proteus vulgaris*, and *Yersinia enterocolitica*) and different species of fungi (*Alternaria alternata*, *Candida guilhermondii*, *Candida parapsilosis*, *Penicillium funiculosum*, *Rhodotorula glutinis*, *Saccharomyces cerevisiae*, *Schizosaccharomyces pombe*, and *Yarrowia lypolytica*).

Table 2. Antimicrobial activity of essential oil from Euphorbiaceae species with occurrence in Caatinga Phytogeographical Domain

Species	Origin	Part plant/Seasons	Major constituents	Microorganism	MIC	Ref.
Croton blanchetianus Baill	Native	Leaves	ND	*Aeromonas hydrophila* *Listeria monocytogenes* *Salmonella Enteritidis*	20 1.25 40	Melo et al. (2013)
Croton campestris A. St.-Hil.	Native	Leaves	β-Caryophyllene Bicyclogermacrene Limonene	*Bacillus cereus* *Escherichia coli* *Pseudomonas aeruginosa* *Staphylococcus aureus* *Shiguella flexineri*	≥ 1024 ≥ 512 ≥ 1024 ≥ 512 ≥ 1024	Almeida et al. (2013)
		Branches	Bicyclogermacrene Spathulenol β-Caryophyllene	*Bacillus cereus* *Escherichia coli* *Pseudomonas aeruginosa* *Staphylococcus aureus* *Shiguella flexineri*	≥ 256 ≥ 512 ≥ 512 ≥ 128 ≥ 512	
Croton grewioides Baill.	Native	Leaves	Estragole Anethole Bicyclogermacrene	*Microsporum canis*** *Candida albicans** *Candida tropicalis**	620/1250 >5000 >2500	Fontenelle et al. (2008)
		Leaves	Not done	*Staphylococcus aureus* *Pseudomonas aeruginosa*	500 > 1000	Rodrigues et al. (2009)
Croton heliotropiifolius Kunth	Native	Leaves	1,8-Cineol 1-Felandreno Cymene	*Aeromonas hydrophila* *Escherichia coli* *Listeria monocytogenes* *Staphylococcus aureus* *Salmonella enteritidis*	10 20 2.5 5 20	Costa et al. (2013)
Croton heliotropiifolius Kunth	Native	Winter	ND	*Serratia marcescens* *Shigella flexineri*	500 >500	Filho et al. (2017)
		Spring	β-Caryophyllene Bicyclogermacrene Germacrene D	*Bacillus cereus* *Enterococcus faecalis* *Escherichia coli* *Klebsiella pneumonia* *Staphylococcus aureus* *Salmonella enterica* *Serratia marcescens* *Shigella flexineri*	> 500 500 500 > 500 > 500 500 500 > 500	

Species	Origin	Part plant/Seasons	Major constituents	Microorganism	MIC	Ref.
Croton limae A.P. Gomes, M.F. Sales P.E. Berry	Native	Leaves	Cedrol Eucalyptol α-Pinene	Candida albicans Candida krusei Candida tropicalis Escherichia coli Klebsiella pneumonia Pseudomonas aeruginosa Staphylococcus aureus	≥1024 ≥1024 ≥1024 ≥1024 ≥1024 ≥1024 512	Leite et al. (2017)
Croton nepetifolius Baill.	Native	Leaves	Methyl-eugenol Bicyclogermacrene β-Caryophyllene	Candida albicans* Candida tropicalis* Microsporum canis*	ND ND >5000	Fontenelle et al. (2008)
Croton tricolor Klotzsch ex Baill.	Nativa	Leaves	Spathulenol Bicyclogermacrene 1,8-Cineol	Candida albicans Candida tropicalis Microsporum canis**	>5000 >5000 9/19	Fontenelle et al. (2008)
		Stems	Epiglobulol α-Bisabolol α-Trans-bergamoto	Candida albicans** Candida tropicalis* Candida krusei**	128/256 256 128/124	Miranda et al. (2019)
Croton heliotropiifolius Kunth	Native	Leaves	Spathulenol 1,8-Cineol o-Cymene	Escherichia coli Pseudomonas aeruginosa Staphylococcus aureus Candida albicans Candida tropicalis	>1024 >1024 >1024 >1024 >1024	Vidal et al. (2016)
		Summer	β-Caryophyllene Bicyclogermacrene 1,8-Cineol	Bacillus cereus Enterococcus faecalis Escherichia coli Klebsiella pneumonia Staphylococcus aureus Salmonella enterica Serratia marcescens Shigella flexneri	500 62.5 500 >500 >500 500 500 500	Filho et al. (2017)
		Autumn	β-Caryophyllene 1,8-Cineol Limonene	Bacillus cereus Enterococcus faecalis Escherichia coli Klebsiella pneumonia Staphylococcus aureus Salmonella enterica Serratia marcescens Shigella flexneri	>500 125 >500 >500 >500 >500 >500 500	

Table 2. (Continued)

Species	Origin	Part plant/Seasons	Major constituents	Microorganism	MIC	Ref.
		Winter	Spathulenol Bicyclogermacrene Germacrene D	Bacillus cereus Enterococcus faecalis Escherichia coli Klebsiella pneumonia Staphylococcus aureus Salmonella enterica	> 500 500 500 > 500 > 500 > 500	
Euphorbia heterophylla L.	Native	Leaves	3,7,12,15-tetramethyl-2-hexadecen-1-ol Octadecanoic acid Oleic acid	Candida albicans Escherichia coli Staphylococcus aureus Streptococcus pneumonia Pseudomonas aeruginosa	25 500 250 250 250	Adedoyin et al. (2013)
		Stems	Octadecanoic acid Oleic acid Cis-cis-Linoleic acid	Candida albicans Escherichia coli Staphylococcus aureus Streptococcus pneumonia Pseudomonas aeruginosa	1000 250 250 500 250	
Jatropha curcas L.	Naturalized	Leaves	δ-Cadinene α-epi-Cadinol Pulegone	Aspergillus niger Bacillus cereus Escherichia coli Klebsiella pneumoniae Micrococcus luteus Pseudomonas aeruginosa Penicillium sp. Rhodotorula rubra Staphylococcus aureus	600 1200 2400 600 300 600 400 300 > 2400	Babahmad et al. (2018)
Jatropha gossypiifolia L.	Native	Leaves	Phytol Pentadecen-5-yne Linalool	Staphylococcus aureus Enterococcus faecium Escherichia coli	100 50 10	Okoh et al. (2016)
		Stem	Phytol Limonene Germacrene D	Staphylococcus aureus Enterococcus faecium Escherichia coli	50 25 100	

Species	Origin	Part plant/Seasons	Major constituents	Microorganism	MIC	Ref.
Ricinus communis L.	Naturalized	Aerial parts	α-Thujone 1,8-Cineol α-Pinene	*Aspergillus niger*	>1024	Zarai et al. (2012)
				Bacillus subtilis	190	
				Bacillus cereus	130	
				Botrytis cinerea	250	
				Escherichia coli	240	
				Enterococcus faecalis	180	
				Enterobacter cloacae	150	
				Fusarium solani	190	
				Klebsiella pneumoniae	320	
				Micrococcus luteus	140	
				Penicillium digitatum	140	
				Pseudomonas aeruginosa	270	
				Staphylococcus aureus	150	
				Staphylococcus epidermidis	120	

* For all strains; ** For different strains; ND – Not Done (without phytochemical identification); MIC – Minimum Inhibitory Concentration (μg/mL); Origin – In relation to Brazil; Major constituents – Organized by order of concentration.

Table 3. Antimicrobial activity of essential oil from Asteraceae species with occurrence in Caatinga Phytogeographical Domain

Species	Origin	Part plant/Seasons	Major constituents	Microorganism	MIC	Ref.
Acanthospermum hispidum DC.	Native	Aerial parts	β-Caryophyllene Bisabolol Germacrene D	*Enterococcus faecalis*** *Lactobacillus plantarum** *Staphylococcus aureus***	>125 >125 125	Alva et al. (2012)
Achyrocline alatan (Kunth) DC.	Native	Aerial parts	Thymol	*Aspergillus flavus* *Aspergillus fumigatus* *Candida parapsilosis* *Candida krusei*	>500 78.7 >500 >500	Zapata et al. (2010)
			Thymol β –Caryophyllene Cymene	*Mycobacterium tuberculosis*	62.5	Bueno-Sánchez et al. (2009)
Acmella ciliate (Kunth) Cass.	Native	Leaves	β-Caryophyllene Germacrene D Caryophyllene oxide	*Candida albicans* *Escherichia coli* *Klebsiella ozaenae* *Proteus mirabilis* *Staphylococcus aureus* *Staphylococcus epidermidis*	>25000 >25000 25000 >25000 15000 15000	Mejía et al. (2012)

Table 3. (Continued)

Species	Origin	Part plant/Seasons	Major constituents	Microorganism	MIC	Ref.
		Flowers	β-Caryophyllene β-Felandreno Caryophyllene oxide	Candida albicans Escherichia coli Proteus mirabilis Klebsiella ozaenae Staphylococcus aureus Staphylococcus epidermidis	25000 15000 25000 25000 15000 15000	
Ageratum conyzoides L.	Native	Leaves	Precoceno I and II Cumarine	Aspergillus flavus	0.01	Nogueira et al. (2010)
Ageratum fastigiatum (Gardner) R.M.King & H.Rob.	Native	Leaves	β-Caryophyllene Germacrene D 1,10-di-epi-cubenol	Escherichia coli Pseudomonas aeruginosa Salmonella typhosa Staphylococcus aureus Streptococcus mutans Streptococcus faecalis	12500 >30000 18000 9000 12500 18000	Del-Vechio-Vieira et al. (2009)
		Branches	Germacrene D β-Caryophyllene 1,10-di-epi-cubenol	Escherichia coli Pseudomonas aeruginosa Salmonella typhosa Staphylococcus aureus Streptococcus mutans Streptococcus faecalis	9000 >30000 6250 5000 9000 12500	
		Roots	β-Caryophyllene α-Humulene Caryophyllene oxide	Escherichia coli Pseudomonas aeruginosa Salmonella typhosa Staphylococcus aureus Streptococcus mutans Streptococcus faecalis	6250 >30000 6250 5000 6250 6250	
Austroeupatorium inulaefolium (Kunth) R.M.King & H.Rob.	Native	Aerial parts	β-Caryophyllene Ledene oxide	Aspergillus flavus ** Aspergillus parasiticus Aspergillus culmorum Fusarium oxysporum Penicillium brevicompactum Penicillium expansum Penicillium nalgiovense	25	Grande-Tovar et al. (2016)

Species	Origin	Part plant/Seasons	Major constituents	Microorganism	MIC	Ref.
Baccharis trinervis (Lam.) Pers.	Native	Aerial Parts	β-Phellandrene Methyl (Z)-dec-2-en-4,6-diynoate Sabinene	Escherichia coli Candida albicans Pseudomonas aeruginosa Staphylococcus aureus Salmomonella chorelaesui	11400 2850 22800 22800 11400	Albuquerque et al. (2004)
			β-Phellandrene (Z)- Lachnophyllum ester Sabinene	Candida albicans ** Candida parapsilosis Candida tropicalis Trichophyton rubrum*	NI NI 150/16310	Sobrinho et al. (2017)
Bidens pilosa L	Naturalized	Leaves	β-Caryophyllene Cadinene Megastigmatrienone	Bacillus cereus Bacillus pumilus Bacillus subtilis Escherichia coli Micrococcus flavus Pseudomonas ovalis	400	Deba et al. (2008)
				Corticium rolfsii Fusarium solani Fusarium oxysporum	100	
		Flowers	Cadinene α-Pinene β-Caryophyllene	Bacillus cereus Bacillus pumilus Bacillus subtilis Escherichia coli Micrococcus flavus Pseudomonus ovalis	400	
				Corticium rolfsii Fusarium solani Fusarium oxysporum	100	
Chromolaena Laevigata (Lam.) R.M.King & H.Rob.	Native	Flowering stage				Murakami et al. (2013)
		Capitula	Laevigatin Germacrene D β-Bisabolene	Candida albicans Escherichia coli Pseudomonas aeruginosa Staphylococcus aureus	1500 1500 500 1500	
		Stems	Spathulenol Laevigatin Furanocadalene	Candida albicans Escherichia coli Pseudomonas aeruginosa Staphylococcus aureus	250 500 500 125	
		Leaves	Laevigatin	Candida albicans	250	

Table 3. (Continued)

Species	Origin	Part plant/Seasons	Major constituents	Microorganism	MIC	Ref.
			Spathulenol	Escherichia coli	500	
			Germacrene D	Pseudomonas aeruginosa	500	
				Staphylococcus aureus	62.5	
		Fruiting stage				
		Cypselas	Laevigatin	Candida albicans	250	
			α-Thujene	Escherichia coli	750	
			Furanocadalene	Pseudomonas aeruginosa	750	
				Staphylococcus aureus	125	
		Stems	Spathulenol	Candida albicans	62,5	
			α-Thujene	Escherichia coli	500	
			Laevigatin	Pseudomonas aeruginosa	500	
				Staphylococcus aureus	62,5	
		Leaves	Laevigatin	Candida albicans	125	
			Spathulenol	Escherichia coli	750	
			α-Thujene	Pseudomonas aeruginosa	500	
				Staphylococcus aureus	62.5	
Chromolaena odoratta (L.) R.M.King & H.Rob.	Native	Leaves	α-Pinene	Aspergillus niger	78	Owolabi et al. (2010)
			β-Pinene	Bacillus cereus	39	
			Germacrene D	Candida albicans	1250	
				Escherichia coli	1250	
				Pseudomonas aeruginosa	1250	
				Staphylococcus aureus	1250	
			β-Guiaene	Aeromonas hydrophila	5000	Prabhu et al. (2011)
			Elemol	Aspergillus flavus Aspergillus	10000	
			Patchoulene	terreus Penicillium notatum	5000	
				Aspergillus niger	> 10000	
				Bacillus cereus	5000	
				Bacillus megaterium	5000	
				Bacillus subtilis	2500	
				Candida albicans	5000	
				Escherichia Coli	2500	
				Pseudomonas auroginosa	2500	
				Salmonella typhi	10000	
				Shigella boydii	2500	

Species	Origin	Part plant/Seasons	Major constituents	Microorganism	MIC	Ref.
Conyza bonariensis (L.) Cronquist	Native	Leaves	Trans-β-farnesene Trans-ocimene β-sesquiphellandrene	Staphylococcus aureus Bacillus cereus Candida albicans Escherichia coli Salmonella typhi Staphylococcus epidermidis	10000 25 100 >100 >100 100	Araujo et al. (2013)
Conyza bonariensis (L.) Cronquist	Native	Leaves	Limonene β-Pinene β-Ocimene	Escherichia coli Salmonella typhi	12.5 6.25	Musembei et al. (2017)
Conyza canadensis (L.) Cronquist	Native	Aerial parts	Limonene Spathulenol β-Pinene	Acinetobacter baumanni Acinetobacter sp Candida albicans Corynebacterium sp. Corynebacterium striatum Escherichia coli Pseudomonas aeruginosa Staphylococcus aureus Staphylococcus epidermidis	0.625 0.625 0.078 0.156 0.3125 0.039 0.3125 0.078 0.3125	Ayaz et al. (2017)
		Roots	cis-lachnophyllum ester (2Z,8Z)-matricaria ester β-Pinene	Acinetobacter baumanni Acinetobacter sp Candida albicans Corynebacterium sp. Corynebacterium striatum Escherichia coli Pseudomonas aeruginosa Staphylococcus aureus Staphylococcus epidermidis	0.625 0.625 0.078 0.625 0.625 0.078 0.625 1.250 0.625	
Eclipta prostrata L.	Native	Leaves	Not done	Bacillus subtillis Escherichia coli Klebsiella pneumonia Rhizopus sp. Salmonella thyphi Staphylococcus aureus	ND	Sureshkumar et al. (2007)
Eclipta prostrata L.	Native	Leaves	Not done	Streptococcus mutans Streptococcus mutans Trichophyton mentographytes	ND	Sureshkumar et al. (2007)

Table 3. (Continued)

Species	Origin	Part plant/Seasons	Major constituents	Microorganism	MIC	Ref.
Egletes viscosa (L)	Native	Flowers	Not done	Trycophyton ruauatii Trycophyton rubrum Staphylococcus aureus**	0.625	Sousa et al. (1998)
Pectis brevipedunculata (Gardner) Sch.Bip.	Native	Aerial parts	Geranial Neral Limonene	Candida albicans Cryptococcus neoformans Escherichia coli Fonsecaea pedrosoi Microsporum canis Microsporum gypseum Staphylococcus aureus Aspergillus niger Staphylococcus epidermidis Trichophyton rubrum	25000 25000 25000 25000 16,666 16,666 16,666 25000 25000 16,666	Marques et al. (2013)
Pectis elongata Kunth	Native	Leaves	Neral Geranial Geranic acid	Aspergillus niger Botrytis cinerea Candida albicans Cylindrocarpon mali Escherichia coli Mycobacterium smegmatis. Pseudomonas aeruginosa Streptococcus faecalis Sclerotinia sclerotiorum Staphylococcus aureus Stereum purpureum	250 125 500 250 500 500 1000 1000 250 500 125	Prudent et al. (1995)
Solidago chilensis Meyen	Native	Leaves	Pumiloxid Limonene γ-Cadinene	Microsporum gypseum Trichophyton mentagrophytes	ND	Vila et al. (2002)
Tanacetum vulgare L.	Naturalized	Flowers	Cis-chrysanthenol Trans-chrysanthenyl acetate Artemiseole	Aliivibrio fischeri Bacillus subtilis Bacillus subtilis Pseudomonas syringae Ralstonia solanacearum Xantomonas euvesicatoria	ND	Móricz, et al. (2015)

Species	Origin	Part plant/Seasons	Major constituents	Microorganism	MIC	Ref.
Tridax procumbens L.	Native	Flowers	(Z)-Falcarinol α-Selinene Limonene	*Aspergillus niger*	60	Joshi, R. K. e al. (2012)
				Aspergillus fumigatus	100	
				Bacillus subtilis	570	
				Enterobacter aerogenes	930	
				Escherichia coli	4580	
				Klebsiella pneumoniae	830	
				Micrococcus flavus	140	
				Micrococcus luteus	260	
				Penicillium chrysogenum	80	
				Proteus mirabilis	670	
				Proteus vulgaris	2080	
				Pseudomonas aeruginosa	ND	
				Salmonella typhimurium	3330	
				Serratia. marcescens	1870	
				Staphylococcus aureus	520	
				Staphylococcus epidermidis	570	
				Streptococcus faecalis	280	

[*] For all strains; [**] For different strains; ND – Not Done (without phytochemical identification); MIC – Minimum Inhibitory Concentration (μg/mL); Origin – In relation to Brazil; Major constituents – Organized by order of concentration.

Table 4. Antimicrobial activity of essential oil from Lamiaceae species with occurrence in Caatinga Phytogeographical Domain

Species	Origin	Part plant/Seasons	Major constituents	Microorganism	MIC	Ref.
Cantinoa mutabilis (Rich.) Harley & J.F.B. Pastore	Native	Leaves	β-Caryophyllene Spathulenol Germacrene D	*Bacillus cereus*	100	Oliva et al. (2006)
				Escherichia coli	Not inhibited	
				Fusarium moniliforme	500	
				Klebsiella spp.	Not inhibited	
				Mucor sp	500	
				Proteus mirabilis	100	
				Pseudomonas aeruginosa	Not inhibited	
				Saccharomyces cerevisiae	500	
				Staphylococcus aureus Staphylococcus epidermidis	100 Not inhibited	

Table 4. (Continued)

Species	Origin	Part plant/Seasons	Major constituents	Microorganism	MIC	Ref.
Leonotis nepetifolia (L.) R.Br.	Naturalized	Aerial parts	(Z)-Phytol Caryophyllene oxide Hexahidrofarnesilacetone	*Bacillus cereus*	25	Casiglia et al. (2014)
				Bacillus subtilis	50	
				Escherichia coli	100	
				Klebsiella pneumonia	100	
				Proteus vulgaris	100	
				Pseudomonas aeruginosa	>100	
				Salmonella typhi	100	
				Staphylococcus aureus	>100	
				Staphylococcus epidermidis	50	
				Streptococcus faecalis	100	
		Leaves	Not done	*Escherichia coli*	10	Gopal et al. (1994)
				Klebsiella pneumonia	10	
				Pseudomonas aeruginosa	Not inhibited	
				Salmonella typhi	10	
				Shigella boydii	Not inhibited	
				Staphylococcus aureus	10	
Leonotis nepetifolia (L.) R.Br.	Naturalized	Leaves	Not done	*Aspergillus niger*	<1	Gopal et al. (1994)
				Aspergillus flavus	<1	
				Aspergillus Fumigatus	<1	
				Microsporum gypseum	1	
				Microsporum nanum	1	
				Trichophyton mentagrophytes	1	
Leonurus japonicus Houtt.	Naturalized	All Parts	Phytone Phytol Caryophyllene Oxide	*Acinetobacter lwoffi*	>3200	Xiong et al. (2013)
				Enterobacter cloacae	>3200	
				Enterococcus faecalis	400	
				Enterococcus faecium	1600	
				Escherichia coli	>3200	
				Klebsiella pneumonia	>3200	
				Macrococcus caseolyticus	200	
				Moraxella catarrhalis	>3200	
				Pseudomonas aeruginosa	>3200	
				Staphylococcus aureus	1600	
				Staphylococcus epidermidis	400	
				Staphylococcus saprophyticus	1600	

Species	Origin	Part plant/Seasons	Major constituents	Microorganism	MIC	Ref.
Martianthus leucocephalus (Mart. ex Benth.) J.F.B. Pastore	Native	Leaves	Carvracol p-Cimene γ-Terpinene	*Bacillus pumillus* *Burkholderia cepacia* *Escherichia coli* *Klebsiella pneumoniae* *Pseudomonas aeruginosa* *Salmonella typhi* *Staphylococcus aureus* *Staphylococcus schleifer*	250 110 50 230 230 110 50 110	Santos et al. (2015)
Medusantha martiusii (Benth.) Harley & J.F.B. Pastore	Native	Leaves	Bicyclogermacrene β-aCryophyllene 1,8-Cineol	*Bacillus cereus* *Escherichia coli*** *Pseudomonas aeruginosa* *Staphylococcus aureus***	256 64/512 512 >1024	Oliveira et al. (2014)
Mentha suaveolens Ehrh.	Naturalized	Aerial parts	Piperitone Oxide p-Cymenol α-Pinene	*Bacillus anthracis* *Candida albicans*** *Candida glabrata* *Citrobacter freundii Enterococcus avium Enterococcus sp* *Escherichia coli* *Klebsiella pneumoniae*** *Proteus mirabilis* *Pseudomonas aeruginosa* *Pseudomonas fluorescens* *Staphylococcus aureus*** *Staphylococcus saprophyticus* *Staphylococcus simulans***	1.69 69 0.69 2.77 2.77 0.69 1.38 0.69 1.38 1.38 13.8 0.69/1.38 1.38	Oumzil, et al. (2002)
Mesosphaerum pectinatum (L.) Kuntze	Native	Leaves	Calamusenone β-Caryophyllene Caryophyllene oxide	*Bacillus subtilis* *Candida albicans* *Candida dubliniensis* *Candida guilhermondii* *Candida parapsilosis* *Cryptococcus neoformans* *Enterococcus faecalis* *Klebsiella pneumoniae* *Pseudomonas aeruginosa Neisseria*	12500 580 3120 12500 25000 1170 50000 200000 200000	Santos et al. (2008)

Table 4. (Continued)

Species	Origin	Part plant/Seasons	Major constituents	Microorganism	MIC	Ref.
			β-Caryophyllene Caryophyllene oxide β-Pinene	gonorrhoeae Salmonella enteritidis Staphylococcus aureus Staphylococcus epidermidis Streptococcus mutans	75000 300000 12500 18750 200000	Nascimento et al. (2008).
Mesosphaerum suaveolens (L.) Kuntze	Native	Aerial parts	Sabinene α-Terpinolene1,8-Cineole	Streptococcus mutans	200	
				Actinomyces pyogenes Escherichia coli Pasteurella multocida Pseudomonas aeruginosa Staphylococcus aureus Streptococcus suis Trychophiton mentagrophytes	1.25 5 1.25 5 0.625 0.625 0.15625	Nantitanon et al. (2007)
Mesosphaerum suaveolens (L.) Kuntze	Native	Aerial parts	4, 11, 11-Trimethyl-8-methylene bicyclo [7.2. 01 undec-4-ene 3-Cyclohexen-1-carboxaldehyde 5-8, 8-8, H,9-8, H, 10 a-Labd-14-ene	Aspergillus niger Bacillus subtilis Candida albicans Escherichia coli Klebsiella pneumoniae Pseudomonas aeruginosa Proteus vulgaris Staphylococcus aureus Trichophyton mentagrophytes Yersinia enterocollitica	40 26 10 26 37 28 15 30 >100 >100	Iwu et al. (1990)
		Leaves	β-Caryophyllene β-Elemene Trans-α-bergamotene	Fusarium moniliforme Mucor sp Saccharomyces cerevisiae	500 500 500	Malele et al. (2003)
Nepeta pérsica Poit. ex Benth	Naturalized	Flowers	4aβ, 7α, 7aβ Nepetalactone 4aα, 7α, 7aβNepetalactone α-Pinene	Enterococcus faecalis Escherichia coli Pseudomonas aeruginosa Salmonella typhi Staphylococcus aureus	50 50 50 50 50	Shafaghat and Oji (2010)

Species	Origin	Part plant/Seasons	Major constituents	Microorganism	MIC	Ref.
Nepeta pérsica Poit. ex Benth	Naturalized	Leaves	4aβ, 7α, 7aβ Nepetalactone 4aα, 7α, 7aβNepetalactone α-Pinene	Enterococcus faecalis Escherichia coli Pseudomonas aeruginosa Salmonella typhi Staphylococcus aureus	50 50 50 50 50	Shafaghat and Oji (2010)
		Stem	4aβ, 7α, 7aβ Nepetalactone 4aα, 7α, 7aβNepetalactone α-Pinene	Enterococcus faecalis Escherichia coli Pseudomonas aeruginosa Salmonella typhi Staphylococcus aureus	50 50 50 50 50	
		Roots	4aβ, 7α, 7aβ Nepetalactone 4aα, 7α, 7aβNepetalactone α-Pinene	Enterococcus faecalis Escherichia coli Pseudomonas aeruginosa Salmonella typhi Staphylococcus aureus	Not inhibited 50 Not inhibited 50 Not inhibited	
Ocimum gratissimum (L.)	Naturalized	Leaves	Not done	Escherichia coli Klebsiella sp Proteus mirabilis Pseudomonas aeruginosa Salmonella enteritidis Shigella flexineri Staphylococcus aureus	6 6 12 >24 3 3 0.75	Nakamura et al. (1999)
			Not done	Aspergillus fumigatus Candida albicans Cryptococcus neoformans Malassezia pachydermatis Microsporum cani Microsporum gypseum Scopulariopsis brevicaulis Trichophyton rubrum Trichophyton interdigitale	>1000 350 300 300 200 150 400 150 250	
Ocimum gratissimum (L.)	Naturalized	Leaves	Not done	Trichophyton mentagrophyte	200	Dubey et al. (2000)
Ocimum campechianum Mill.	Native	Leaves	Eugenol β-Caryophyllene Bicyclogermacrene	Candida glabrata Candida tropicalis Candida krusei Candida parapsilosis** Candida albicans**	625 1250 1250 312,5 1250	Vieira et al. (2014)

Table 4. (Continued)

Species	Origin	Part plant/Seasons	Major constituents	Microorganism	MIC	Ref.
		Leaves	Eugenol β-Caryophyllene α-Elemene	Enterococcus faecalis Escherichia coli Pseudomonas aeruginosa Rhodotorula glutinis Schizosaccharomyces cerevisiae Schizosaccharomyces pombe Staphylococcus aureus Yarrowia lypolytica	104 35 173 139 69 104 104 69	Sachetti (2004)
Vitex agnus-castu (L.)	Naturalized	Immature fruit	Sabinene 1,8-Cineol α-Pinene	Alternaria alternata Aspergillus flavus Aspergillus niger Aspergillus ochraceus Bacillus subtilis Escherichia coli Fusarium tricinctum Micrococcus flavus	44.5 178.0 178.0 178.0 445.0 219.0 178.0 445.0	Stojkovic et al. (2011)
			Sabinene 1,8-Cineol α-Pinene	Penicillium funiculosum Penicillium ochrochloron Salmonella typhimurium Staphylococcus aureus Trichoderma viride	178.0 130.0 44.5 219.0 219.0	Stojkovic et al. (2011)
		Mature fruit	1,8-Cineol Sabinene α-Pinene	Alternaria alternata Aspergillus flavus Aspergillus niger Aspergillus ochraceus Bacillus subtilis Escherichia coli Fusarium tricinctum Micrococcus flavus Penicillium funiculosum Penicillium ochrochloron Salmonella typhimurium Staphylococcus aureus Trichoderma viride	89.0 219.0 219.0 219.0 890.0 219.0 89.0 445.0 178.0 219.0 44.5 219.0 219.0	

Species	Origin	Part plant/Seasons	Major constituents	Microorganism	MIC	Ref.
		Leaves	1,8-Cineol α-Pinene Trans-b-arnesene	Alternaria alternata Aspergillus flavus Aspergillus niger Aspergillus ochraceus Bacillus subtilis Escherichia coli Fusarium tricinctum Micrococcus flavus Penicillium funiculosum	130.0 178.0 178.0 219.0 890.0 219.0 178.0 445.0 178.0	
Vitex agnus-castu (L.)	Naturalized	Leaves	1,8-Cineol α-Pinene Trans-b-arnesene	Penicillium ochrochloron Salmonella typhimurium Staphylococcus aureus Trichoderma viride	178.0 44.5 219.0 267.0	Stojković et al. (2011)
		Seeds	1,8-Cineol Sabinene α-Pinene	Candida albicans Candida dubliniensis Candida glabrata Candida krusei Candida parapsilosis Candida lusitaniae Candida famata Candida tropicalis	30 270 270 270 1060 2130 2130 130	Asdadi et al. (2015)
		Leaves at 8 h	cis-Calamenene 6,9-guaiadiene Caryophyllene oxide	Candida albicans Candida krusei Candida parapsilosis Candida tropicalis Trichophyton rubrum	2500 > 5000 > 5000 > 5000 70	Pereira et al. (2018)
Vitex gardneriana Schauer	Native	Leaves at 12 h	Caryophyllene oxide cis-Calamenene 6,9-guaiadiene	Candida albicans Candida krusei Candida parapsilosis Candida tropicalis Trichophyton rubrum	> 5000 > 5000 > 5000 2500 70/150	
		Leaves at 17h	cis-Calamenene 6,9-guaiadiene Caryophyllene oxide	Candida albicans Candida krusei Candida parapsilosis Candida tropicalis Trichophyton rubrum	2500 2500 2500 2500 70	

* For all strains; ** For different strains; ND – Not Done (without phytochemical identification); MIC – Minimum Inhibitory Concentration (µg/mL); Origin – In relation to Brazil; Major constituents – Organized by order of concentration.

Generally, the main phytochemical compounds determine the biological properties of EOs (Bilia et al. 2014). Concerning these components, the family Euphorbiaceae presents bicyclegermacrene, 1,8-cineol, β-Caryophyllene, spathulenol, and limonene as recurrent compounds among the studied species. The family Asteraceae presents Germacrene D, Limonene, and β-Caryophyllene and the family Lamiaceae presents β-Caryophyllene, 1,8 cieol, and α-pinene.

In this sense, it is clear that Lamiaceae species family possess more effectiveness against microorganism tested, including the Gram-negative bacteria, when compared with the Euphorbiaceae and Asteraceae family. Probably the metabolites present in these two families are not able to penetrate the double membrane of Gram-negative bacteria (Holley and Patel 2005).

There are several mechanism of action in the microorganisms by the EOs (Kalemba and Kunicka 2003). The rupture or alteration of the bacterial membrane is the most effective mode of action (Burt 2004). Such activity can be seen through 1,8-cineol by membrane permeability alteration. And when associated with other antimicrobial agents, can facilitate the actions causing lethal damage (Carson et al. 2002; De Souza et al. 2010; Rao et al. 2010). Studies with α-pinene also demonstrated inhibition activity in 50% of phospholipases from *Cryptococcus neoformans* at subinhibitory concentrations (Silva et al. 2012). It is conjectured that these properties against fungi and bacteria are related to decreased cytosolic pH, inhibition of calcium channels, H-ATPase pumps, and breakage in the electron transport chain (Tangarlin et al. 1999), as well as alteration in the cytoplasmic membrane.

The physicochemical characteristics of the EOs facilitate various biological properties. It is considered that lipophilic structural properties of the EOs increased membrane's permeability leading to extravasation of the cytoplasmic content (Solórzano-Santos and Miranda-Novales, 2012). However, the lipophilic sesquiterpenoid hydrocarbon Germacrene D not act in the membrane permeability in the differents pathogenic microorganisms (Deuschle et al. 2007). Thus, it is suggested that Germacrene D has a mechanism of action distinct from sesquiterpenes, and may act in the proteins of cell membranes (Cowan 1999) or interact with hydrophobic protein moieties (Juven et al. 1994; Sikkema et al. 1995).

Another role of the EOs can be observed in species with high concentrations of Limonene. First described by Brennan et al. (2013), this compound caused perturbations in the cell wall, induced changes in the compensatory response to cell wall damage and demonstrated activity at the gene level against *Saccharomyces cerevisiae*. Thus, this monoterpene has great potential when used in studies of human pathogens (Singh et al. 2010; Espina et al. 2013).

The β-caryophyllene and biclygermacrene, present in the family Euphorbiaceae were considered as having low to moderate activity in bacteria and fungi (Del-Vechio-Vieira et al. 2009). Both are natural sesquiterpenes found abundantly in essential oils (Costantin et al. 2001; Cysne et al. 2005, Deba et al. 2008; Hussain et al. 2008). However, the composition of the minor metabolites of the EOs contributes to the antimicrobial activity (Burt 2004).

Therefore, the antimicrobial potential of the essential oils of plant species that occurred in the Caatinga was evidenced. These species showed considerable antibacterial and antifungal activities against several species of microorganisms, including multi-drug resistant pathogens. Moreover, 1,8-cineol and α-pinene are probably responsible for the observed antimicrobial activities. However, β-Caryophyllene was also present in the species of all families studied, but until now its mechanism of action has not been elucidated. Unfortunately, studies of the antimicrobial activity of the EOs from species that occur in CPD are still scarce. On the other hand, as shown in this study, there is a potential of the species that occur in CPD and therefore, tests that analise the antimicrobial efficacy of the other native species of this region are recommended.

REFERENCES

Ab'Saber, Aziz. (1997). "The morphoclimatic domains in South America." *Geomorfologia.*, (52) 1-22.

Albuquerque, Maria Rose Jane R., et al. (2004). "Composition and antimicrobial activity of the essential oil from aerial parts of *Baccharis trinervis* (Lam.) Pers." *ARKIVOC.*, 6, 59-65.

Albuquerque, Ulysses Paulino., et al. (2012). Caatinga revisited: ecology and conservation of an important seasonal dry forest. *The Scientific World Journal.*, 2012, PMC3415163. Doi: 10.1100/2012/205182.

Almeida, Thiago Silva., et al. (2013). "Chemical composition, antibacterial and antibiotic modulatory effect of *Croton campestris* essential oils." *Industrial Crops and Products.*, 44, 630–633. Doi:10.1016/j.indcrop.2012.09.010.

Alva, Mariana., et al. (2012). "Bioactivity of the Essential Oil of an Argentine Collection of *Acanthospermum hispidum* (Asteraceae)." *Natural Product Communications.*, 7(2), 245-248. Doi: 10.1177/1934578X1200700235.

Araújo, Floricéa Magalhães., et al. (2017). "Antibacterial activity and chemical composition of the essential oil of *Croton heliotropiifolius* Kunth from Amargosa, Bahia, Brazil." *Industrial crops and products.*, 105, 203-206. Doi: 10.1016/j.indcrop.2017.05.016.

Araujo, Liliana., et al. (2013). "Chemical Composition and Biological Activity of *Conyza bonariensis* Essential Oil Collected in Mérida, Venezuela." *Natural Product Communications.*, 8(8), 1934578X 1300800. Doi: 10.1177/1934578X1300800838

Arcoverde, José Hélton Vasconcelos., et al. (2014). "Screening of Caatinga plants as sources of lectins and trypsin inhibitors." *Natural Product Research.*, 28, 1297–1301. Doi: 10.1080/14786419.2014.900497.

Asdadi, Ali., et al. (2015). "Study on chemical analysis, antioxidant and in vitro antifungal activities of essential oil from wild *Vitex agnus*-castus L. seeds growing in area of Argan Tree of Morocco against clinical strains of *Candida* responsible for nosocomial infections." *Journal de Mycologie Médicale.*, 25(4), 118–127. Doi: 10.1016/j.mycmed.2015.10.005.

Ayaz, Fatma., Küçükboyacı, Nurgün. & Demirci, Betül. (2017). "Chemical composition and antimicrobial activity of the essential oil of *Conyza canadensis* (L.) Cronquist from Turkey." *Journal of Essential Oil Research.*, 29(4), 336–343. Doi: 10.1080/10412905.2017.1279989.

Babahmad, Rachid Ait., et al. (2018). "Chemical composition of essential oil of *Jatropha curcas* L. leaves and its antioxidant and antimicrobial activities." *Industrial Crops and Products.*, 121, 405–410. Doi: 10.1016/j.indcrop.2018.05.030.

Bakkali, Fadil., et al. (2008). "Biological effects of essential oils-A review." *Food and Chemical Toxicology.*, 46, 446–475. Doi: 10.1016/j.fct.2007.09.106.

Bilia, Anna Rita., et al. (2014). "Essential oils loaded in nanosystems: A developing strategy for a successful therapeutic approach." *Evid. Based Complement. Altern. Med.*, 2014, 1-14. Doi: 10.1155/2014/651593.

Brennan, Timothy C. R., Krömer, Jens. O. & Nielsen, Lars Keld. (2013). "Physiological and Transcriptional Responses of Saccharomyces cerevisiae tod-Limonene Show Changes to the Cell Wall but Not to the Plasma Membrane." *Applied and Environmental Microbiology.*, 79(12), 3590–3600. Doi: 10.1128/AEM.00463-13.

Bueno-Sanchez, Juan Gabriel., et al. (2009). "Anti-tubercular activity of eleven aromatic and medicinal plants occurring in Colombia." *Biomédica.*, 29(1), 51-60. Doi: 10.7705/biomedica.v29i1.41.

Burt, Sara. (2004). Essential Oils: "Their Antibacterial Properties and Potential Applications in Foods- A Review." *International Journal of Food Microbiology.*, 94(3), 223-53. Doi: 10.1016/ j.ijfoodmicro.2 004.03.022.

Calo, Juliany Rivera., et al. (2015). "Essential oils as antimicrobials in food systems–A review." *Food Control.*, 54, 111-119. Doi: 10.1016/j.food cont.2014.12.040.

Carson, Christine Frances., Mee, Brian J. & Riley, Thomas V. (2002). "Mechanism of action of *Melaleuca alternifolia* (tea tree) oil on *Staphylococcus aureus* determined by time-kill, lysis, leakage, and salt tolerance assays and electron microscopy." *Antimicrobial Agents and Chemotherapy.*, 46, 1914–1920. Doi: 10.1128/aac.46.6.1914-1920.20 02.

Casiglia, Simona., Bruno, Maurizio. & Senatore, Felice. (2014). "Activity against Microorganisms Affecting Cellulosic Objects of the Volatile Constituents of *Leonotis nepetaefolia* from Nicaragua." *Natural Product Communications.*, *9*(11), 1637-9. Doi: 10.1177/ 1934578X140 0901127.

Costa, Ana Caroliny Vieira., et al. (2013). "Chemical composition and antibacterial activity of essential oil of a *Croton rhamnifolioides* leaves Pax & Hoffm." *Semina: Ciências Agrárias.*, *34*(6), 2853-2864. Doi: 10.5433/1679-0359.2013v34n6p2853.

Costantin, Mara B., et al. (2001). "Essential oils from *Piper cernuum* and *Piper regnellii*: antimicrobial activities and analysis by GC/MS and 13C-NMR." *Planta medica.*, *67*(08), 771-773. Doi: 10.1055/s-2001-18363.

Cysne, Juliana de Brito., et al. (2005). "Leaf essential oils of four *Piper* species from the State of Ceará - Northeast of Brazil." *Journal of the Brazilian Chemical Society.*, *16*(6), 1378-138. Doi: 10.1590/S0103-50532005000800012.

De Souza, Evandro Leite., et al. (2010). "Influence of *Origanum vulgare* L. essential oil on enterotoxin production, membrane permeability Bioactivity of Essential Oils Towards Fungi and Bacteria: Mode of Action and Mathematical Tools 243 and surface characteristics of *Staphylococcus aureus*." *International Journal of Food Microbiology.* 137: 308–311. Doi: 10.1016/j.ijfoodmicro.2009.11.025.

Deba, Farah., et al. (2008). "Chemical composition and antioxidant, antibacterial and antifungal activities of the essential oils from *Bidens pilosa* Linn. var. Radiata." *Food control.*, *19*(4), 346-352. Doi: 10.1016/j.foodcont.2007.04.011.

Del-Vechio-Vieira, Glauciemar., et al. (2009). "Chemical Composition and Antimicrobial Activity of the Essential Oils of *Ageratum fastigiatum* (Asteraceae). *Academy of Chemistry of Globe Publications.*," 15 (4), 258-263.

Deuschle, Régis Augusto Norbet., et al. (2007). "Fractionation of Senecio desiderabilis Vellozo dichloromethane extract and evaluation of antimicrobial activity." *Revista Brasileira de Farmacognosia.*, *17*(2), 220-223. Doi: 10.1590/S0102-695X2007000200015.

Dima, Cristian. & Dima, Stefan. (2015). "Essential oils in foods: extraction, stabilization, and toxicity." *Current Opinion in Food Science.*, *5*, 29-35. Doi: 10.1016/j.cofs.2015.07.003.

Dubey, Nawal Kishore., et al. (2000). "Antifungal properties of *Ocimum gratissimum* essential oil (ethyl cinnamate chemotype)." *Fitoterapia.* 71(5): 567–569. Doi:10.1016/S0367-326X(00)00206-9.

Eslahi, Hassan., Fahimi, Nafiseh. & Sardarian, Ali Reza. (2018). Chemical Composition of Essential Oils Hassan, in Hashemi, SM (ed.) Essential Oils in Food Processing - Chemistry, Safety and Applications. John Wiley & Sons., pp. 119-171.

Espina, Laura., et al. (2013). "Mechanism of bacterial inactivation by (+)-limonene and its potential use in food preservation combined processes." *PloS one.*, *8*(2), e56769. Doi: 10.1371/journal.pone.0056769.

Filho, José Marcos Teixeira A., et al. (2017). "Chemical composition and antibacterial activity of essential oil from leaves of *Croton heliotropiifolius* in different seasons of the year." *Revista Brasileira de Farmacognosia.*, *27*(4), 440–444. Doi: 10.1016/j.bjp.2017.02.004.

Flora do Brazil 2020 in construction. Botanical Garden From Rio de Janeiro. Disponível em: < http://floradobrasil.jbrj.gov.br/ >. Access in: 14 Jul. 2019.

Fonseca, Maira Christina Marques., et al. (2015). "Potential of medicinal plant essential oils in the control of plant pathogens." *Revista Brasileira de Plantas Medicinais.*, *17*(1), 45-50.

Fontenelle, Raquel Oliveira dos Santos., et al. (2008). "Antifungal activity of essential oils of Croton species from the Brazilian Caatinga biome." *Journal of Applied Microbiology.*, *104*(5), 1383–1390.

Forzza, Rafaela Campostrini., et al. (2013). Introduction. In: Species list from Brazilian Garden. Botanical Garden from Rio de Janeiro.

Gopal, R. Hamsaveni., Vasanth, Saradha. & Vasudevan, S. V. (1994). "Antimicrobial activity of essential oil of *Leonotis nepetaefolia.*" *Ancient Science of Life.*, *14*(1-2), 68–70.

Grande-Tovar, Carlos David. et al. (2016). "Sub-lethal concentrations of Colombian *Austroeupatorium inulifolium* (H.B.K.) essential oil and its effect on fungal growth and the production of enzymes." *Industrial Crops and Products.*, *87*, 315–323. Doi:10.1016/ j.indcrop.20 16.04.066.

Holley, Richard A. & Patel, Dhaval. (2004). "Improvement in shelf-life and safety of perishable foods by plant essential oils and smoke antimicrobials." *Food Microbiol.*, *22*, 273-292. Doi: 10.1016/ j.fm.2004.08.006.

Hussain, Abdullah Ijaz., et al. (2008). "Chemical composition, antioxidant and antimicrobial activities of basil (*Ocimum basilicum*) essential oils depends on seasonal variations." *Food chemistry.*, *108*(3), 986-995. Doi: 10.1016/j.foodchem.2007.12.010.

IBGE - Instituto Brasileiro de Geografia e Estatística. (2004). "Mapa de Biomas do Brasil, primeira aproximação". Rio de Janeiro: IBGE. Acessible at <www.ibge.gov.br>.

Iwu, Maurice Mmaduakolam., et al. (1990). "Antimicrobial Activity and Terpenoids of the Essential Oil of *Hyptis suaveolens*" *International Journal of Crude Drug Research.*, *28*(1), 73–76. Doi:10.3109/ 13880209009082783.

Jayasena, Dinesh. & Jo, Cheorun. (2013). "Essential oils as potential antimicrobial agents in meat and meat products: A review." *Trends in Food Science & Technology.*, *34*(2), 96-108. Doi:10.1016/ j.tifs.2013.09.002.

Joshi, Rajesh. & Badakar, Vijaylaxmi. (2012). "Chemical Composition and *in vitro* Antimicrobial Activity of the Essential Oil of the Flowers of *Tridax procumbens*." *Natural Product Communications.*, *7*(7), 1-7. Doi:10.1177/1934578X1200700736.

Juven, B. J., et al. (1994). "Factors that interact with the antibacterial action of thyme essential oil and its active constituents." *Journal of Applied Bacteriology.*, *76*, 626-631. Doi:10.1111/j.1365-2672.1994.tb01661.x.

Kalemba, Danuta. & Kunicka-Styczńska, Alina. (2003). "Antibacterial and Antifungal Properties of Essential Oils." *Current Medicinal Chemistry.*, *10*, 813-829. Doi:10.2174/0929867033457719.

Langeveld, Wendy., Veldhuizen, Edwin. & Burt, Sara. (2014). "Synergy between essential oil components and antibiotics: a review" *Critical reviews in microbiology.*, *40*(1), 76-94. Doi:10.3109/1040841X.20 13.763219.

Leal, Inara Roberta., et al. (2005). "Changing the course of biodiversity conservation in the Caatinga of northeastern Brazil". *Conservation Biology.*, *19*(3), 701-706. Doi: 10.1111/j.1523-1739.2005.00703.x.

Leite, Tiago Rodrigues., et al. (2017). "Antimicrobial, modulatory and chemical analysis of the oil of *Croton limae*." *Pharmaceutical Biology.*, *55*(1), 2015–2019. Doi:10.1080/13880209.2017.1355926.

Malele, R. S., et al. (2003). "Essential Oil of *Hyptis suaveolens* (L.) Poit. from Tanzania: Composition and Antifungal Activity." *Journal of Essential Oil Research.*, *15*(6): 438–440. Doi:10.1080/10412905.2 003.9698633.

Marques, Andre Mesquita., et al. (2013). "Traditional use, chemical composition and antimicrobial activity of *Pectis brevipedunculata* essential oil: A correlated lemongrass species in Brazil." *Emirates Journal of Food and Agriculture.*, *25*(10), 771-798. Doi:10.9755/ejfa .v25i10.16408.

Mejía, Carlos Andrés Rincón., Osorio, Jhon Carlos Castaño. & Vázquez, Eunice Ríos. (2012). "Biological activity of essential oils from *Acmella ciliata* (Kunth) Cass". *Revista Cubana de Plantas Medicinales*, *17*(2), 160-171.

Melo, Geiseanny Fernandes do Amarante., et al. (2013). "The sensitivity of bacterial foodborne pathogens to *Croton blanchetianus* Baill essential oil." *Brazilian Journal of Microbiology.*, *44*(4), 1189–1194. Doi:1 0.1590/S1517-83822014005000009.

Miranda, Fabrício Mendes., et al. (2018). "Promising antifungal activity of *Croton tricolor* stem essential oil against *Candida* yeasts." *Journal of Essential Oil Research.*, 1–5. Doi:10.1080/10412905.2018.1539416.

Móricz, Ágnes., et al. (2015). "Tracking and identification of antibacterial components in the essential oil of *Tanacetum vulgare* L. by the combination of high-performance thin-layer chromatography with direct bioautography and mass spectrometry." *Journal of Chromatography A.*, *142*, 310–317. Doi:10.1016/j.chroma.2015 .10.0 10.

Moura, Patrícia., et al. (2015). "Caatinga plants as weapons against microorganisms: advances and challenges in Méndez-Vilas A. (Ed) The Battle Against Microbial Pathogens: Basic Science, Technological Advances and Educational ProGrams" *FORMATEX Microbiology Series.*

Murakami, Cynthia., et al. (2013). "Chemical Composition and Antimicrobial Activity of Essential Oils from *Chromolaena laevigata* during Flowering and Fruiting Stages." *Chemistry & Biodiversity.*, *10*(4), 621–627. Doi:10.1002/cbdv.201200025.

Musembei, Racheal. & Joyce, Kiplimo Jepkorir. (2017). "Chemical Composition and Antibacterial Activity of Essential Oil from Kenyan *Conyza bonariensis* (L.) *Cronquist."* *Science Letters.*, *5*(2), 180-185.

Nakamura, Celso Vataru., et al. (1999). "Antibacterial activity of *Ocimum gratissimum* L. essential oil." *Memórias Do Instituto Oswaldo Cruz.* 94(5), 675–678. Doi:10.1590/S0074-02761999000500022.

Nantitanon, Witayapan., howwanapoonpohn, Sombat. & Okonogi, Siriporn. (2007). "Antioxidant and Antimicrobial Activities of *Hyptis suaveolens* Essential Oil." *Scientia Pharmaceutica.*, 75(1), 35–46. Doi:10.3797/ scipharm.2007.75.35.

Nascimento, P. F. C., et al. (2008). "*Hyptis pectinata* essential oil: chemical composition and anti-Streptococcus mutans activity." *Oral Diseases.* 14(6), 485–489. Doi:10.1111/j.1601-0825.2007.01405.x.

Nogueira, Juliana Hellmeister de campos., et al. (2010). "Ageratum conyzoides essential oil as aflatoxin suppressor of *Aspergillus flavus*." *International Journal of Food Microbiology.*, *137*(1), 55–60. Doi:10.1111/j.1601-0825.2007.01405.x.

Okoh, Sunday., et al. (2016). "Antibacterial and Antioxidant Properties of the Leaves and Stem Essential Oils of *Jatropha gossypifolia* L. *BioMed Research International.*," 1–9. Doi:10.1155/2016/9392716.

Oliva, Maria M., et al. (2006). "Antimicrobial Activity and Composition of *Hyptis mutabilis* Essential Oil." *Journal of Herbs, Spices & Medicinal Plants.*, *11*(4), 57–63. Doi: 10.1300/J044v11n04_07.

Oliveira, Allan Demetrius Leite., et al. (2014). "Menezes IR. Chemical Composition, Modulatory Bacterial Resistance and Antimicrobial Activity of Essential Oil the *Hyptis martiusii* Benth by Direct and Gaseous Contact." *Jundishapur Journal of Natural Pharmaceutical Products.*, *9*(3), e13521.

Olson, David., et al. (2001). "Terrestrial Ecoregions of the World: A New Map of Life on EarthA new global map of terrestrial ecoregions provides an innovative tool for conserving biodiversity." *BioScience.*, *51*(11), 933-938.

Oumzil, H., et al. (2002). "Antibacterial and antifungal activity of essential oils of *Mentha suaveolens*." *Phytotherapy Research.*, *16*(8), 727–731. Doi:10.1002/ptr.1045.

Owolabi, Moses., et al. (2010). "Chemical Composition and Bioactivity of the Essential Oil of *Chromolaena odorata* from Nigeria." *Records of Natural Products.*, *4*(1), 72-78.

Pereira, Evaristo José Pires., et al. (2018). "Circadian Rhythm, and Antimicrobial and Anticholinesterase Activities of Essential Oils from *Vitex gardneriana*." *Natural Product Communications.*, *13*(5), 1934578X1801300. Doi:10.1177/1934578X1801300528.

Prabhu, Velliangiri., et al. (2011). "Essential oil composition, antimicrobial, MRSA and *in-vitro* cytotoxic activity of fresh leaves of *Chromolaena odorata*." *Journal of Pharmacy Research.*, *4*(12), 4609-4611.

Prudent, D., et al. (1995). "Analysis of the Essential Oil of *Pectis elongata* Kunth. from Martinique. Evaluation of Its Bacteriostatic and Fungistatic Properties." *Journal of Essential Oil Research.*, *7*(1), 63–68. Doi:10.1080/10412905.1995.9698464.

Rao, Anjana., et al. (2010). "Mechanism of antifungal activity of terpenoid phenols resembles calcium stress and inhibition of the TOR pathway" *Antimicrobial Agents and Chemotherapy.*, 54, 5062–5069. Doi:10.1128/AAC.01050-10.

Raut, Jayant Shankar. & Karuppayil, Sankunny Mohan. (2014). "A status review on the medicinal properties of essential oils." *Industrial crops and products.*, *62*, 250-264. Doi:10.1016/j.indcrop.2014.05.055.

Rodrigues, Fabíola., Costa, José. & Coutinho, Henrique. (2009). "Synergy effects of the antibiotics gentamicin and the essential oil of *Croton zehntneri*" *Phytomedicine.*, *16*(11), 1052–1055. Doi:10.1016/j.phymed.2009.04.004.

Roohinejad, Shahin., et al. (2018). "Extraction Methods of Essential Oils From Herbs and Spices in Hashemi, SM (ed.) Essential Oils in Food Processing - Chemistry, Safety and Applications." John Wiley & Sons., pp. 119-171.

Santos, Patrícia Oliveira., et al. (2008). "Chemical composition and antimicrobial activity of the essential oil of *Hyptis pectinata* (L.) Poit." *Química Nova.*, *31*(7), 1648–1652. Doi:10.1590/S0100-40422008000700009.

Santos, Suikinai Nobre., et al. (2015). "Chemical composition and antibacterial activity of the essential oil of *Hyptis leucocephala*." *Asa, São Paulo.*, *3*(2), 3-11.

Savoia, Dianella. (2012). "Plant-derived antimicrobial compounds: alternatives to antibiotics." *Future microbiology.*, *7*(8), 979-990. Doi:10.2217/fmb.12.68.

Shafaghat, Ali. & Oji, Kodhamali. (2010). "Nepetalactone Content and Antibacterial Activity of the Essential Oils from Different Parts of *Nepeta persica*." *Natural Product Communications.*, *5*(4), 1934578X1000500. Doi:10.1177/1934578X1000500427.

Shojaee-Aliabadi, Saeedeh., Hosseini, Seyed Mohammad Bagher. & Mirmoghtadaie, Liela. (2018). Antimicrobial Activity of Essential Oil in Hashemi, SM (ed.) Essential Oils in Food Processing - Chemistry, Safety and Applications. John Wiley & Sons., pp. 119-171.

Sikkema, Jan., De Bont, Jan. & Poolman, Bert. (1995). "Mechanisms of membrane toxicity of hydrocarbons." *Microbiological Reviews.*, *59*, 201-222

Silva, Ana Cristina Rivas., et al. (2012). "Biological activities of a-pinene and β-pinene enantiomers." *Molecules.*, *17*(6), 6305-6316. Doi: 10.3390/molecules17066305.

Singh, Priyanka., et al. (2010). "Chemical profile, antifungal, antiaflatoxigenic and antioxidant activity of *Citrus maxima* Burm. and *Citrus sinensis* (L.) Osbeck essential oils and their cyclic monoterpene, DL-limonene." *Food and Chemical Toxicology.*, *48*(6), 1734-1740. Doi:10.1016/j.fct.2010.04.001.

Sobrinho, Antonio Carlos Nogueira., et al. (2016). "Chemical composition, antioxidant, antifungal and hemolytic activities of essential oil from *Baccharis trinervis* (Lam.) Pers. (Asteraceae)." *Industrial Crops and Products.*, *84*, 108–115. Doi:10.1016/j.indcrop.2016.01.051.

Solórzano-Santos, Fortino. & Miranda-Novales, Maria Guadalupe. (2012). "Essential oils from aromatic herbs as antimicrobial agents." *Current Opinion in Biotechnology.*, *23*, 136-141. Doi:10.1016/ j.copbio.2011. 08.005.

Stojković, Dejan., et al. (2011). "Chemical composition and antimicrobial activity of *Vitex agnus*-castus L. fruits and leaves essential oils." *Food Chemistry.*, *128*(4), 1017–1022.

Sureshkumar, Shanmugam., et al. (2007). "Antimicrobiological Studies on Different Essential Oils of *Wedelia* Species (W. chinensis, W. trilobata and W. biflora) and *Eclipta alba* (Asteraceae)." *Asian Journal of Chemistry.*, *19*(6), 4674-4678.

Swamy, Mallappa Kumara., Akhtar, Mohd., Sayeed, Sinniah. & Uma, Rani. (2016). "Antimicrobial properties of plant essential oils against human pathogens and their mode of action: an updated review." *Evidence-Based Complementary and Alternative Medicine*, Article ID 3012462, 21 pages. Doi:10.1155/2016/3012462.

Tangarlin, José Renato., et al. (1999). "Medicinal plants and plant pathogen control." *Biotecnol. Cien. Desenv.*, *2*(1), 16-22.

Vidal, Cinara Soares., et al. (2016). "Chemical composition, antibacterial and modulatory action of the essential oil of *Croton rhamnifolioides* leaves pax and Hoffman." *Bioscience Journal.*, *32*, 1632-1643. Doi:10.14393/BJ-v32n1a2016-33918.

Vieira, Priscila R. N., et al. (2014). "Chemical composition and antifungal activity of essential oils from *Ocimum* species." *Industrial Crops and Products.*, *55*, 267–271. Doi:10.1016/j.indcrop.2014.02.032.

Vila, Roser., et al. (2002). "Composition and Antifungal Activity of the Essential Oil of *Solidago chilensis*." *Planta Medica.*, *68*(2), 164–167. Doi:10.1055/s-2002-20253.

Xiong, Liang., et al. (2013). "Chemical Composition and Antibacterial Activity of Essential Oils from Different Parts of *Leonurus japonicus* Houtt." *Molecules.*, *18*(1), 963–973. Doi:10.3390/molecules18010963.

Zapata, Bibiana., et al. (2010). "Antifungal and cytotoxic activity of essential oils of plants of the Asteraceae family." *Revista Iberoamericana de Micología.*, *27*(2), 101–103. Doi:10.1016/j.riam.2010.01.005.

Zarai, Zied., et al. (2012). "Essential oil of the leaves of Ricinus communis L.: *In vitro* cytotoxicity and antimicrobial properties." *Lipids in Health and Disease.*, *11*(1), 102. Doi: 10.1186/1476-511X-11-102.

In: Antimicrobial Potential of Essential Oils   ISBN: 978-1-53616-945-4
Editors: B. Oliveira de Veras et al.   © 2020 Nova Science Publishers, Inc.

Chapter 6

ANTIBACTERIAL POTENTIAL OF ESSENTIAL OIL FROM *SYZYGIUM AROMATICUM* (L.) MERR. AND L. M. PERRY

João Ricardhis Saturnino de Oliveira[1,],*
Cristiane Marinho Uchôa Lopes[2],
Rebeca Xavier da Cunha[1],
Francisco Henrique da Silva[1],
Bruno Oliveira de Veras[1],
José Galberto Martins da Costa[3],
Vera Cristina Oliveira de Carvalho[1]
and Vera Lúcia de Menezes Lima[1]

[1]Department of Biochemistry, Universidade Federal de Pernambuco, Recife, Pernambuco, Brazil
[2]Department of Morphology, Universidade Federal do Cariri, Barbalha, Ceará, Juazeiro do Norte - CE, Brazil
[3]Department of Biological Chemistry, Regional University of Cariri, Crato, Ceará, Brazil

---

[*] Corresponding Author's Email: ricardhis@gmail.com.

Abstract

Syzygium aromaticum is used in culinary, but this herb is also used for anti-inflammatory and antimicrobial purposes. The aim of this study was to investigate reports of *S. aromaticum* essential oil antimicrobial activity. Literature review was conducted in several databases, and data from essential oil related to bacteria strains were collected. There are reports of *S. aromaticum* essential oil against 52 strains of bacteria; while Eugenol was tested against 33. S. aromaticum killed all strains tested, but had better results against *Vibrio choleare*. Thus, *S. aromaticum* essential oil might be used as a natural tool against bacteria.

Introduction

Herbs form a group of plants with wide spectrum for culinary usage (Reichling et al. 2009; Bertella et al. 2018). Moreover, the search for other possibilities of usage and harnessing for these plants, such as rosemary, cumin and cinnamon, have found pharmacological and biotechnological potential (Ponce et al. 2003; Reichling et al. 2009; Santoro et al. 2007). From these activities, antioxidant (Costa et al. 2012; Bakour et al. 2018; Naveed et al. 2013), anti-inflammatory (Bakour et al. 2018; Foddai et al. 2019), and antimicrobial have been reported in extracts and essential oils of these plants (Císarová et al. 2016; Deans et al. 1995).

Another well-known herb that has been studied throughout the last two decades is clove (*Syzygium aromaticum*) (Cortés-Rojas, de Souza and Oliveira 2014). Flowers, buds and leaves have been studied, and several reports indicate that clove has extra usability, for example, as a potent antioxidant (Bakour et al. 2018; Deans et al. 1995; Radünz et al. 2019; Gülçin, Elmastaş, and Aboul-Enein 2012), due to secondary metabolites found in extracts and essential oil. Thus, some companies are using additives from clove to help food storage (Kovács et al. 2016; Lee et al. 2009; Lin et al. 2019).

Due to its analgesic effect, essential oil is being used, also, to help during odontological procedures (Zhang et al. 2017; Moon, Kim, and Cha 2011; Alqareer, Alyahya and Andersson 2006).

Figure 1. Representation of clove flowers and dry buds.

Besides that, several studies have proved that eugenol, the major constituent of *S. aromaticum* essential oil, is the active compound among the volatile species found in clove essential oil (Bicas et al. 2011; Machado et al. 2011; Kheawfu et al. 2018). Eugenol, 4-alil-2-metoxiphenol, is being associated with several activities due to its capability to cross cell membrane (Kovács et al. 2016; Naveed et al. 2013; Lee et al. 2009; Kheawfu et al. 2018; Chen et al. 2017), and clove is known to be the best source of it (Bhuiyan 2012; Lee et al. 2009; Chen et al. 2017).

Antimicrobial potential of essential oils is highly possible, because of their easy mechanism to cross or break cell membranes, the alteration in cell permeability, and antioxidant activity inside cells (Hammer, Carson and Riley 1999; Zhang et al. 2017; Essid et al. 2017; Chen et al. 2017). And, natural products are a pharmacological class with lesser biological risk than synthetic to side effects besides higher chance to combat resistant bacterias (Naveed et al. 2013; Biba et al. 2014; Souza et al. 2016). Although literature reports antimicrobial activity of clove compounds (Nishigaki et al. 2010; Reichling et al. 2009), products with it on its constitution are not produced to worldwide usage as antibiotics. In fact, most cases of clove usage are found in popular medicine (Cortés-Rojas, de Souza and Oliveira 2014) and through scientific reports (Moon, Kim and Cha 2011; Naveed et al. 2013). This study aimed to reunite evidences of antimicrobial activity of Syzygium aromaticum essential oil and to elucidate the possibility of its applications as a commercial product.

METHODS

Strategy of Review

This review of literature had an integrative approach. Investigation was conducted through the databases Science Direct, Cochrane Library, SciELO, PubMed, Google Scholar, and Lilacs. Keywords used are listed as following: *Syzygium aromaticum*, clove, essential oil, volatile compounds, antimicrobial activity, antimicrobial potential, biofilm, bacteria, microorganisms, and antibiotic. Keywords were also translated to Portuguese, Spanish, French, and Italian in order to increase sensibility. Only reports of essential oils from clove were accepted. Data from other species from *Syzygium* were excluded. Studies with eugenol from other plants were not accepted. There was not restriction of date, language, or source of information. Origin of strains were described as reported in corresponding studies.

RESULTS AND DISCUSSION

Syzygium aromaticum essential oil (SaEO) was tested against 52 different strains; and eugenol, its major compound, against 33, as listed in Table 1.

Table 1. List of bacterias tested for antibacterial activity of essential oil *Syzygium aromaticum* and eugenol

Species and Origin	Essential oil	Eugenol
Acinetobacter baumanii	0.25*[4]	-
Acinetobacter calcoacetica NCIB 8250	7.9**[1] 10.3±0.2**[2]	15.4±0.3**[2]
Aeromonas hydrophila NCTC 8049	11.5**[1] 11.7±1.1**[2]	17.0±0.4**[2]
Alcaligenes faecalis NCIB 8156	9.8**[1] 23.1±0.6**[2]	12.3±0.5**[2]
Bacilus licheniformis	7.6±2.8*[5]	19.3±0.33*[5]
Bacilus subtilis	19.5±0.5**[3] 0.125*[3]	-
Bacilus subtilis NCIB 3610	9.0**[1] 21.1±0.1**[2]	21.8±0.4**[2]
Beneckea natriegens ATCC 14048	8.5**[1] 15.8±0.7**[2]	20.8±1.8**[2]
Brevibacterium linens NCIB 8456	10.0**[1] 29.8±0.1**[2]	12.7±0.1**[2]
Brocothrix thermosphacta	9.6**[1] 11.1±0.1**[2]	14.1±0.2**[2]
Citrobacter freundii NCIB 11490	10.0**[1] 14.1±2.6**[2]	9.1±0.3**[2]
Clostridium perfringens NCIB 10696	9.0**[1] 13.4±0.5**[2]	9.7±0.1**[2]
Enterococcus faecalis	0.5*[4]	-
Enterococcus faecalis NCTC 775	15.5±0.6**[2]	10.0±0.1**[2]
Enterobacter aerogenes	14.2±0.75**[6]	-
Enterobacter aerogenes NCTC 10006	7.6**[1] 7.8±1.1**[2]	9.9±0.1**[2]
Erwinia carotovora NCPPB 312	20.4**[1] 11.7±0.4**[2]	10.0±0.1**[2]
Escherechia coli	16.3±1.3**[3] 0.125*[3] 0.25*[4] 5.4±1.08*[5] 11.87±3.22**[6]	31.6±0.88*[5]
Escherechia coli O157:H7	15.7±1.5	-
Escherechia coli NCIB 8879	10.0**[1] 13.6±0.3**[2]	13.3±0.2**[2]
Flavobacterium suaveolens NCIB 8992	8.8**[1] 14.4±0.2**[2]	11.6±0.6**[2]

Table 1. (Continued)

Species and Origin	Essential oil	Eugenol
Klebisiella ozaenae	14.5±2.5[**6]	-
Klebsiella pneumoniae	0.25[*4] 12.0±3.15[**6]	-
Klebsiella pneumoniae NCIB 418	7.7[**1] 9.1±0.1[**2]	10.9±0.3[**2]
Lactobacillus plantarum NCDO 343	9.7[**1] 28.5±1.0[**2]	21.5±0.6[**2]
Leuconostoc cremoris NCDO 543	8.4[**1] 18.7±0.6[**2]	Resistant[2]
Micrococcus luteus NCIB 8165	8.6[**1] 14.8±0.8[**2]	11.7±0.7[**2]
Moraxella sp. NCIB 10762	8.9[**1] 15.8±0.8[**2]	10.1±0.6[**2]
Proteus mirabilis	16.5±0.5	-
Proteus vulgaris	18.2±1.3[**3] 0.125[*3]	-
Proteus vulgaris NCIB 4175	7.3[**1] 9.1±0.6[**2]	8.3±0.3[**2]
Pseudomonas aeruginosa	9.5±0.5[**3] 0.5[*3] >2.0[*4] 18.86±1.46[**6]	-
Pseudomonas aeruginosa NCIB 950	4.0[**1] 14.0±1.9[**2]	15.5±0.6[**2]
Pseudomonas fluorescens	8.6±2.1[*5]	Resistant[*5]
Salmonella paratyphi	4.3±1.08[*5]	22.3±0.88[*5]
Salmonella pullorum NCTC 10704	9.4[**1] 14.0±0.8[**2]	12.9±0.1[**2]
Salmonella typhi	18.0±3.08[**6]	-
Salmonella typhi D1	5.4±1.08[*5]	24.6±0.66[*5]
Salmonella typhi G7	3.26±0.05[*5]	21.0±0.57[*5]
Salmonella typhimurium	>2.0[*4]	-
Serratia marcescens	0.25[*4] 14.25±0.43[**6]	-
Serratia marcescens NCIB 1377	8.8[**1] 21.6±0.9[**2]	22.9±0.8[**2]
Shigella dysenteriae	16.5±0.5[**6]	-
Staphylococcus aureus	16.3±0.7[**3] 0.125[*3] 0.25[*4]	22.6±0.88[*5]

Species and Origin	Essential oil	Eugenol
	5.4±1.08*[5]	
Staphylococcus aureus NCIB 1377	7.2**[1]	-
Staphylococcus aureus NCIB 6571	14.9±0.1**[2]	11.5±0.5**[2]
Staphylococcus aureus	0.187*[7]	0.211*[7]
Staphylococcus epidermidis	16.8±1.2**[3] 0.250*[3]	-
Streptococus faecalis NCTC 775	4.0**[1]	-
Vibrio cholerae	23.75±3.03**[6]	-
Yersinia enterocolotica NCTC 10460	8.4**[1] 13.7**[2]	11.6±0.4**[2]
Listeria monocytogenes	>9.0**[1]	-

[1]Dean et al. 1995; [2]Dorman, Deans 2000; [3]Fu et al. 2007; [4]Hammet et al. 1999; [5]Naveed et al. 2013; [6]Saeed, Tariq 2008; [7]Budri et al. 2015; *Minimun inibitory concentration (%v/v), **Diameter of inhibition zone (mm). Note that some studies do not report the origin of strains.

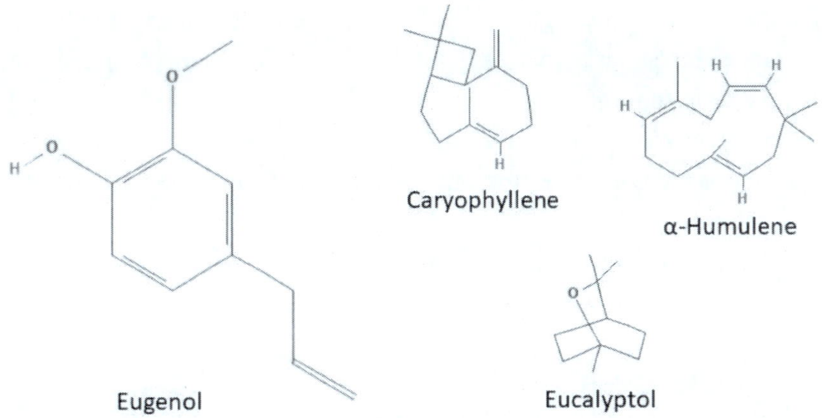

Caryophyllene

α-Humulene

Eugenol

Eucalyptol

Figure 2. Major constituents of Syzygium aromaticum essential oil.

SaEO combated both gram types, positive and negative, from all strains evaluated (Budri et al. 2015; Deans et al. 1995; Dorman and Deans 2000; Saeed and Tariq 2008; Hammer, Carson and Riley 1999; Fu 2007). However, eugenol did not stop growth, neither killed, *Leuconostoc cremoris* (NCDO 543) and *Pseudomonas fluorescens* (Dorman and Deans 2000). All studies used essential oil from leaves of clove, and reported high percentage of eugenol in their constitution. Other major compounds were Caryophyllene oxide, Eucalyptol and α-Humulene, but these had concentrations below 15%, much lesser than eugenol, with reports up to 90% (Budri et al. 2015; Deans et al. 1995; Dorman and Deans 2000; Hammer, Carson, and Riley 1999; Naveed et al. 2013; Fu 2007; Saeed and Tariq 2008).

From the 52 strains tested, only 30 were tested twice (Deans et al. 1995; Dorman and Deans 2000; Fu 2007; Hammer, Carson, and Riley 1999). In addition, some studies did not report the identification of strains, which may be different on each study.

Besides that, some studies used minimum inhibitory concentration method to assess antibacterial activity (Fu 2007; Hammer, Carson and Riley 1999; Naveed et al. 2013), while others used diameter of inhibition zone to do so (Dorman and Deans 2000; Saeed and Tariq 2008; Budri et al. 2015).

These two methodologies, actually, are often performed together to figure out whether the compounds are bacteriostatics or bactericides, and to identify better dosages.

When used separately, it is difficult to compare both results, because one methodology completes the other (Hammer, Carson and Riley 1999; Brennan-Krohn and Kirby 2019).

SaEO was highly reported to attack bacteria through alterations in membrane permeability, cell lysis, inhibiting bacteria growth, and altering bacteria metabolism, leading to death (Kovács et al. 2016; Saeed and Tariq 2008; Deans et al. 1995).

These mechanisms are also related to eugenol activity (Dorman and Deans 2000; Naveed et al. 2013; Budri et al. 2015). However, according to literature evidences, the other volatile compounds found in SaEO are also good against bacteria (Budri et al. 2015; Dorman and Deans 2000; Fu 2007; Hammer, Carson, and Riley 1999; Naveed et al. 2013; Saeed and Tariq 2008; Verma et al. 2016; Costa et al. 2011), indicating a higher activity due to all major components.

That may, too, respond why SaEO was efficient against *Leuconostoc cremoris* (NCDO 543) and *Pseudomonas fluorescens*, and eugenol was not (Naveed et al. 2013).

CONCLUSION

Literature heavily supports *Syzygium aromaticum* essential oil antibacterial potential. Eugenol, its major compound, is very effective, but components found in lesser concentrations also play important roles. Although other studies regarding its toxicity, and interaction with the human body are needed, clove essential oil might be a great source for antibacterial medicines.

REFERENCES

Alqareer, Athbi, Asma Alyahya and Lars Andersson. 2006. "The Effect of Clove and Benzocaine versus Placebo as Topical Anesthetics". *Journal of Dentistry,* 34 (10): 747 - 50. https://doi.org/10.1016/j.jdent.20 06.01. 009.

Bakour, Meryem, Najoua Soulo, Nawal Hammas, Hinde El Fatemi, Abderrazak Aboulghazi, Amal Taroq, Abdelfattah Abdellaoui, Noori Al-Waili and Badiaa Lyoussi. 2018. "The Antioxidant Content and Protective Effect of Argan Oil and Syzygium Aromaticum Essential Oil in Hydrogen Peroxide-Induced Biochemical and Histological Changes". *International Journal of Molecular Sciences,* 19 (2). https://doi.org/10.3390/ijms19020610.

Bertella, Anis, Kheira Benlahcen, Sidaoui Abouamama, Diana C. G. A. Pinto, Karim Maamar, Mebrouk Kihal and Artur M. S. Silva. 2018. "Artemisia Herba-Alba Asso. Essential Oil Antibacterial Activity and Acute Toxicity". *Industrial Crops and Products,* 116 (October 2017): 137 - 43. https://doi.org/10.1016/j.indcrop.2018.02.064.

Bhuiyan, Mohammad Nazrul Islam. 2012. "Constituents of the Essential Oil from Leaves and Buds of Clove (Syzigium Caryophyllatum (L.) Alston)". *African Journal of Pharmacy and Pharmacology,* 6 (16). https://doi.org/10.5897/ajpp10.004.

Biba, V. S., A. Amily, S. Sangeetha and P. Remani. 2014. "Anticancer, Antioxidant and Antimicrobial Activity of *Annonaceae* Family".

World Journal of Pharmacy and Pharmaceutical Sciences, 3 (3): 1595 - 1604.

Bicas, Juliano Lemos, Gustavo Molina, Ana Paula Dionísio, Francisco Fábio Cavalcante Barros, Roger Wagner, Mário Roberto Maróstica and Gláucia Maria Pastore. 2011. "Volatile Constituents of Exotic Fruits from Brazil". *Food Research International*, 44 (7): 1843 - 55. https://doi.org/10.1016/j.foodres.2011.01.012.

Brennan-Krohn, Thea and James E. Kirby. 2019. "When One Drug Is Not Enough: Context, Methodology, and Future Prospects in Antibacterial Synergy Testing". *Clinics in Laboratory Medicine*. Elsevier Inc. https://doi.org/10.1016/j.cll.2019.04.002.

Budri, P. E., N. C. C. Silva, E. C. R. Bonsaglia, A. Fernandes, J. P. Araújo, J. T. Doyama, J. L. Gonçalves, M. V. Santos, D. Fitzgerald-Hughes and V. L. M. Rall. 2015. "Effect of Essential Oils of Syzygium Aromaticum and Cinnamomum Zeylanicum and Their Major Components on Biofilm Production in Staphylococcus Aureus Strains Isolated from Milk of Cows with Mastitis". *Journal of Dairy Science*, 98 (9): 5899 - 5904. https://doi.org/10.3168/jds.2015-9442.

Chen, Xiangning, Lupei Ren, Menglin Li, Jia Qian, Junfeng Fan and Bin Du. 2017. "Effects of Clove Essential Oil and Eugenol on Quality and Browning Control of Fresh-Cut Lettuce". *Food Chemistry*, 214: 432 - 39. https://doi.org/10.1016/j.foodchem.2016.07.101.

Císarová, Miroslava, Dana Tančinová, Juraj Medo and Miroslava Kačániová. 2016. "The in Vitro Effect of Selected Essential Oils on the Growth and Mycotoxin Production of Aspergillus Species". *Journal of Environmental Science and Health - Part B Pesticides, Food Contaminants, and Agricultural Wastes*, 51 (10): 668 - 74. https://doi.org/10.1080/03601234.2016.1191887.

Cortés-Rojas, Diego Francisco, Claudia Regina Fernandes de Souza and Wanderley Pereira Oliveira. 2014. "Clove (Syzygium Aromaticum): A Precious Spice". *Asian Pacific Journal of Tropical Biomedicine*, 4 (2): 90 - 96. https://doi.org/10.1016/S2221-1691(14)60215-X.

Costa, Emmanoel Vilaça, Lívia Macedo Dutra, Hugo César Ramos de Jesus, Paulo Cesar de Lima Nogueira, Valéria Regina de Souza

Moraes, Marcos José Salvador, Sócrates Cabral de Holanda Cavalcanti, Roseli La Corte dos Santos and Ana Paula do Nacimento Prata. 2011. "Chemical Composition and Antioxidant, Antimicrobial, and Larvicidal Activities of the Essential Oils of Annona Salzmannii and A. Pickelii (*Annonaceae*)". *Natural Product Communications*, 6(6): 1934578X1100600. https://doi.org/10.1177/1934578x11006 0063 6.

Costa, Emmanoel Vilaça, Lívia Macedo Dutra, Paulo Cesar de Lima Nogueira, Valéria Regina de Souza Moraes, Marcos José Salvador, Luis Henrique Gonzaga Ribeiro and Fernanda Ramos Gadelha. 2012. "Essential Oil from the Leaves of Annona Vepretorum : Chemical Composition and Bioactivity". *Natural Product Communications*, 7(2): 1934578X1200700. https://doi.org/10.1177/1934578x1200700240.

Deans, S. G., R. C. Noble, R. Hiltunen, W. Wuryani and L. G. Pénzes. 1995. "Antimicrobial and Antioxidant Properties of Syzygium Aromaticum (L.) Merr. and Perry: Impact upon Bacteria, Fungi and Fatty Acid Levels in Ageing Mice". *Flavour and Fragrance Journal*, 10 (5): 323 - 28. https://doi.org/10.1002/ffj.2730100507.

Dorman, H. J. D. and Stanley G. Deans. 2000. "Antimicrobial Agents from Plants: Antibacterial Activity of Plant Volatile Oils". *Journal of Applied Microbiology*, 88 (2): 308 - 16. https://doi.org/10.1046/j.1365-2672.2000.00969.x.

Essid, Rym, Majdi Hammami, Dorra Gharbi, Ines Karkouch, Thouraya Ben Hamouda, Salem Elkahoui, Ferid Limam and Olfa Tabbene. 2017. "Antifungal Mechanism of the Combination of Cinnamomum Verum and Pelargonium Graveolens Essential Oils with Fluconazole against Pathogenic Candida Strains". *Applied Microbiology and Biotechnology*, 101 (18): 6993 - 7006. https://doi.org/10.1007/s00253-017-8442-y.

Foddai, Marzia, Mauro Marchetti, Alessandro Ruggero, Claudia Juliano and Marianna Usai. 2019. "Evaluation of Chemical Composition and Anti-Inflammatory, Antioxidant, Antibacterial Activity of Essential Oil of Sardinian Santolina Corsica Jord. and Fourr". *Saudi Journal of*

Biological Sciences, 26 (5): 930 - 37. https://doi.org/10.1016/ j.sjbs. 2018.08.001.

Fu, Y.; 2007. "Antimicrobial Activity of Clove and Rosemary Essential Oils Alone and in Combination". Phytotherapy Research, 21 (1): 989 - 94. https://doi.org/10.1002/ptr.2179.

Gülçin, Ilhami, Mahfuz Elmastaş and Hassan Y. Aboul-Enein. 2012. "Antioxidant Activity of Clove Oil - A Powerful Antioxidant Source". Arabian Journal of Chemistry, 5 (4): 489 - 99. https://doi.org/10.1016/ j.arabjc.2010.09.016.

Hammer, K. A., C. F. Carson and T. V. Riley. 1999. "Antimicrobial Activity of Essential Oils and Other Plant Extracts". Journal of Applied Microbiology, 86 (6): 985 - 90. https://doi.org/10.1046/j.1365-2672.1999.00780.x.

Kheawfu, Kantaporn, Surachai Pikulkaew, Thomas Rades, Anette Müllertz and Siriporn Okonogi. 2018. "Development and Characterization of Clove Oil Nanoemulsions and Self-Microemulsifying Drug Delivery Systems". Journal of Drug Delivery Science and Technology, 46 (May): 330 - 38. https://doi.org/10.1016/j.jddst.2018.05.028.

Kovács, J. K., H. Ábrahám, A. Böszörményi, P. Felső, L. Makszin, T. Palkovics, Z. Pápai, G. Schneider, G. Horváth and L. Emődy. 2016. "Antimicrobial and Virulence-Modulating Effects of Clove Essential Oil on the Foodborne Pathogen Campylobacter Jejuni". Applied and Environmental Microbiology, 82 (20): 6158 - 66. https://doi.org/ 10.1128/aem.01221-16.

Lee, Seongwei, Musa Najiah, Wee Wendy and Musa Nadirah. 2009. "Chemical Composition and Antimicrobial Activity of the Essential Oil of Syzygium Aromaticum Flower Bud (Clove) against Fish Systemic Bacteria Isolated from Aquaculture Sites". Frontiers of Agriculture in China, 3 (3): 332 - 36. https://doi.org/10.1007/s11703-009-0052-8.

Lin, Lin, Xuefang Mao, Yanhui Sun, Govindan Rajivgandhi and Haiying Cui. 2019. "Antibacterial Properties of Nanofibers Containing Chrysanthemum Essential Oil and Their Application as Beef

Packaging". *International Journal of Food Microbiology*, 292 (May 2018): 21 - 30. https://doi.org/10.1016/j.ijfoodmicro.2018.12.007.

Machado, M., A. M. Dinis, L. Salgueiro, José B. A. Custódio, C. Cavaleiro and M. C. Sousa. 2011. "Anti-Giardia Activity of Syzygium Aromaticum Essential Oil and Eugenol: Effects on Growth, Viability, Adherence and Ultrastructure". *Experimental Parasitology*, 127 (4): 732 - 39. https://doi.org/10.1016/j.exppara.2011.01.011.

Moon, Sang Eun, Hye Young Kim and Jeong Dan Cha. 2011. "Synergistic Effect between Clove Oil and Its Major Compounds and Antibiotics against Oral Bacteria". *Archives of Oral Biology*, 56 (9): 907 - 16. https://doi.org/10.1016/j.archoralbio.2011.02.005.

Naveed, Rasheeha, Iftikhar Hussain, Abdul Tawab, Muhammad Tariq, Moazur Rahman, Sohail Hameed, M. Shahid Mahmood, Abu Baker Siddique and Mazhar Iqbal. 2013. "Antimicrobial Activity of the Bioactive Components of Essential Oils from Pakistani Spices against Salmonella and Other Multi-Drug Resistant Bacteria". *BMC Complementary and Alternative Medicine*, 13 (1): 1. https://doi.org/10.1186/1472-6882-13-265.

Nishigaki, Ikuo, Rajendran Peramaiyan, Venugopal Ramachandran, Ekambaram Gnapathy, Sakthisekaran Dhanapal and Nishigaki Yutaka. 2010. "Cytoprotective Role of Astaxanthin against Glycated Protein/Iron Chelate-Induced Toxicity in Human Umbilical Vein Endothelial Cells". *Phytotherapy Research*, 24 (June): 54 - 59. https://doi.org/10.1002/ptr.

Ponce, A. G., R. Fritz, C. Del Valle and S. I. Roura. 2003. "Antimicrobial Activity of Essential Oils on the Native Microflora of Organic Swiss Chard". *LWT - Food Science and Technology*, 36 (7): 679 - 84. https://doi.org/10.1016/S0023-6438(03)00088-4.

Radünz, Marjana, Maria Luiza Martins da Trindade, Taiane Mota Camargo, André Luiz Radünz, Caroline Dellinghausen Borges, Eliezer Avila Gandra and Elizabete Helbig. 2019. "Antimicrobial and Antioxidant Activity of Unencapsulated and Encapsulated Clove (Syzygium Aromaticum, L.) Essential Oil". *Food Chemistry*, 276: 180 - 86. https://doi.org/10.1016/j.foodchem.2018.09.173.

Reichling, Jürgen, Paul Schnitzler, Ulrike Suschke and Reinhard Saller. 2009. "Essential Oils of Aromatic Plants with Antibacterial, Antifungal, Antiviral, and Cytotoxic Properties - An Overview". *Forschende Komplementarmedizin*, 16 (2): 79 - 90. https://doi.org/10.1159/000207196.

Saeed, Sabahat and Perween Tariq. 2008. "In Vitro Antibacterial Activity of Clove against Gram Negative Bacteria". *Pakistan Journal of Botany*, 40 (5): 2157 - 60.

Santoro, Giani F., Maria G. Cardoso, Luiz Gustavo L. Guimarães, Lidiany Z. Mendonça and Maurilio J. Soares. 2007. "Trypanosoma Cruzi: Activity of Essential Oils from Achillea Millefolium L., Syzygium Aromaticum L. and Ocimum Basilicum L. on Epimastigotes and Trypomastigotes". *Experimental Parasitology*, 116 (3): 283 - 90. https://doi.org/10.1016/j.exppara.2007.01.018.

Souza, Heloisa, Cícera Fernandes, Saulo Tintino, Maria Morais-Braga, Antonia Araújo, Henrique Coutinho, Irwin Menezes and Marta Kerntopf. 2016. "Phytochemical Composition, Antibacterial and Modulatory of Antibiotic Activity of the Extract and Fractions from Annona Squamosa L.". *Ethnobiology and Conservation*, 2 (9): 1 - 8. https://doi.org/10.15451/ec2013-8-2.9-1-08.

Verma, Ram Swaroop, Neeta Joshi, Rajendra Chandra Padalia, Ved Ram Singh, Prakash Goswami and Amit Chauhan. 2016. "Characterization of the Leaf Essential Oil Composition of Annona Squamosa L. from Foothills of North India". *Medicinal and Aromatic Plants*, 5 (5). https://doi.org/10.4172/2167-0412.1000270.

Zhang, Yi, Yue Wang, Xiaojing Zhu, Ping Cao, Shaomin Wei and Yanhua Lu. 2017. "Antibacterial and Antibiofilm Activities of Eugenol from Essential Oil of Syzygium Aromaticum (L.) Merr. and L. M. Perry (Clove) Leaf against Periodontal Pathogen Porphyromonas Gingivalis". *Microbial Pathogenesis*, 113 (September): 396 - 402. https://doi.org/10.1016/j.micpath.2017.10.054.

In: Antimicrobial Potential of Essential Oils   ISBN: 978-1-53616-945-4
Editors: B. Oliveira de Veras et al.   © 2020 Nova Science Publishers, Inc.

Chapter 7

ANTIBACTERIAL ACTIVITY OF ESSENTIAL OILS FROM SPECIES OF *ANNONA* L.

João Ricardhis Saturnino de Oliveira[1,*],
Weber Melo Nascimento[1],
Ana Paula Sant'Anna da Silva[1],
Vera Cristina Oliveira de Carvalho[1],
Bruno Oliveira de Veras[1], *Bianka Santana dos Santos*[2]
and Vera Lúcia de Menezes Lima[1]

[1]Department of Biochemistry, Universidade Federal de Pernambuco, Recife, Pernambuco, Brazil
[2]Life Sciences Center, Universidade Federal de Pernambuco, Caruaru, Pernambuco, Brazil

ABSTRACT

Annona is one of the most important genus from *Annonaceae*. Several studies indicate activity and new biocompounds found in extracts from leaves, barks, and fruits from the genus, although little is known about essential oil possibilities. This study aimed to review antibacterial

---

[*] Corresponding Author's Email: ricardhis@gmail.com.

activity of species from the genus *Annona*. This review searched antibacterial studies from all Annona species, without excluding date of publication, language, and strains. Eight species of Annona were reported against several strains. *Bacillus* and *Staphylococcus* species were the most tested, and Annonas had the better results for these strains. Studies diverge method of antibacterial activity and strains, which turns difficult to compare essential oils capabilities. *A. squamosa* and *A. cherimola* were the most tested, and had the highest potential.

INTRODUCTION

Annona is one of the most known genus of *Annonaceae* family (Biba et al. 2014). Although *Annonaceae* has over 100 genera, this genus is one of the most studied (Rabelo et al. 2015), because fruits from this genus are well received in culinary, due to their sweetness and high concentration of nutritional compounds (Bicas et al. 2011). These fruits, such as custard apple and soursop, are also a source for economics, because of their market value (Rabelo et al. 2015). However, once pulp is extracted from fruits, the rest is often discharged (Biba et al. 2014; Chen et al. 2017; Neethu, Santhoshkumar, and Kumar 2016).

In order to solve this problem, studies have been conducted to evaluate new usage for these plants (Bicas et al. 2011; Chavan, Shinde, and Nirmal 2006; El-Chaghaby, Ahmad, and Ramis 2014; Javed et al. 2013). Thus, the obtention of essential oils from parts of plants from the genus Annona became common worldwide; and, have been already reported to have antioxidant, anti-inflammatory, antitumor, and antidiabetic effects (Rabelo et al. 2015).

Besides that, several countries seek for essential oils with antimicrobial activity (Biba et al. 2014). With the increase of multiresistant bacteria, and the need of natural conservants to increase food warranty, essential oils are a promising source to help food and health industries (Javed et al. 2013; Neethu, Santhoshkumar, and Kumar 2016; Padhi et al. 2011; Lin et al. 2019; do Evangelho et al. 2019). Antioxidant potential of Annona species essential oils are heavily reported in the literature, and this is one glimpse of antimicrobial capability (Costa et al. 2011; El-Chaghaby, Ahmad, and

Ramis 2014; Rabelo et al. 2015), but little is known about the antimicrobial potential of essential oils from these plants (Rabelo et al. 2015).

This review aimed to collect reports of antibacterial activity of essential oils from Annona species. With the purpose of reunite published data regarding this subject, and compare what species had higher efficiency, and what strains were most vulnerable.

METHODS

Strategy of Review

This is an integrative review of literature. The following keywords were used to search information regarding the subject: Annona, *Annonaceae*, essential oil, volatile compounds, antimicrobial activity, antimicrobial potential, antibacterial, biofilm, bacteria, microorganisms, and antibiotic. These key words were also translated to Portuguese, French, Spanish, and Italian, for searches conducted in google scholar. Research was conducted in the databases, ScienceDirect, Cochrane Library, PubMed, SciELO, Google Scholar, and Lilacs. All reports of antibacterial activity of essential oils from Annona species were accepted. There was not restriction of date, language, or source of information.

RESULTS AND DISCUSSION

Annona has approximately 162 species (Rabelo et al. 2015). Several studies reported antibacterial activity from Annona species, but most of them tested compounds found in extracts (Javed et al. 2013; Padhi et al. 2011; Biba, Jeba Malar, and Remani 2013; El-Chaghaby, Ahmad, and Ramis 2014; Ruddaraju et al. 2019; Pinto et al. 2017; Rinaldi et al. 2017). Thus, eight species of Annona are commented in this study due to evidence of antimicrobial activity of their essential oils being found in literature.

Essential oils from *Annona cherimola* (Elhawary et al. 2013; Mohammed et al. 2016), *Annona glabra* (Elhawary et al. 2013), *Annona muricata* (Elhawary et al. 2013), *Annna pickelii* (Costa et al. 2011), *Anona salzmannii* (Costa et al. 2011), *Annona squamosa* (Elhawary et al. 2013; Mohammed et al. 2016), *Annona vepretorum* (Costa et al. 2012), and *Annona cherimola* x *Annona squamosa* (hybrid) (Mohammed et al. 2016) presented antimicrobial activity against several different strains, as listed in Table 1. All studies obtained essential oils from leaves.

The major compounds found in essential oils differ from each specie: Annona cherimola essential oil was 25.02% β-Elemene, 17.71% Germacrene-D, 9.50% β-Caryophyllene, and 6.30% Bicyclogermacrene, (Elhawary et al. 2013), but also reported with 17.06% α-Copaene (Mohammed et al. 2016). *Annona glabra* had 37.11% β-Caryophyllene, 19.19% γ-Murulene, and 11.12% α-Humulene (Elhawary et al. 2013). *Annona muricata* had 23.58% Bicycloelemene, 16.85% Limonene and 14.30% β-Pipene (Elhawary et al. 2013). *Annna pickelii* 45.40% Bicyclogermacrene, 14.60% ε-Caryophyllene and 10.60% α-Copaene (Costa et al. 2011). *Anona salzmannii* had 20.3% Bicyclogermacrene, 19.90% ε-Caryophyllene and 15.30% δ-Cadinene (Costa et al. 2012). *Annona squamosa* had 42.49% β-Gurjumene, 6.68% Viridiflorene and 5.72% γ-Murulene (Elhawary et al. 2013), but also had 27.59% Isocaryophyllene (Mohammed et al. 2016). *Annona vepretorum* had 43.70% Bicyclogermacrene, 11.40% Spathulenol, and 10.0% α-Felandrene (Costa et al. 2012).

Surprisingly, *Annona cherimola* x *Annona squamosa* (hybrid) had 21.78% α-Copaene, 13.99% Caryophyllene and 4.63% α-Elemene (Mohammed et al. 2016), these reports indicate different percentages than both species used in the creation of this hybrid, suggesting that its metabolism for volatile components is probably different from both mother species.

Compositions of essential oils presented higher concentrations of Bicyclogermacrene, Bicycloelemene, β-Caryophyllene, β-Elemene, β-Gurjumene, Germacrene-D, ε-Caryophyllene, Isocaryophyllene, and α-Copaene. These compounds have been mentioned already in literature as antimicrobial. Their activity was associated with alterations in membrane permeability and in electrons transport through membrane, besides cell destruction (Wang et al. 2017; Hu et al. 2019; Foddai et al. 2019; Salazar et al. 2018).

Figure 1. Species from the genus Annona with essential oils tested for antimicrobial activity up to 2019.

Table 1. List of Annona species with essential oils tested for antibacterial activity

Species	Strains	Effect
Annona cherimola (Elhawary et al. 2013; Mohammed et al. 2016)	*Bacillus cereu*[2]	35*
	Bacillus subtilis[2]	28*
	Bacillus subtilis (ATCC 6051)[1]	15*
	Escherichia coli (ATCC 11775)[1]	14*
	Pseudomonas aeruginosa[2]	30*
	Pseudomonas aeruginosa (ATCC 10145)[1]	16*
	Staphylococcus aureus[2]	26*
	Staphylococcus aureus (ATCC 12600)[1]	14*

Table 1. (Continued)

Species	Strains	Effect
Annona glabra (Elhaway et al. 2013)	Bacillus subtilis (ATCC 6051)	19*
	Escherichia coli (ATCC 11775)	22*
	Pseudomonas aeruginosa (ATCC 10145)	23*
	Staphylococcus aureus (ATCC 12600)	17*
Annona muricata (Elhawary et al. 2013)	Bacillus subtilis (ATCC 6051)	18*
Annona muricata (Elhawary et al. 2013)	Escherichia coli (ATCC 11775)	19*
	Pseudomonas aeruginosa (ATCC 10145)	20*
	Staphylococcus aureus (ATCC 12600)	15*
Annona pickelii (Costa et al. 2011)	Escherichia coli (ATCC 10538)	Resistant
	Pseudomonas aeruginosa (ATCC27853)	Resistant
	Staphylococcus aureus (ATCC6538)	Resistant
	Staphylococcus aureus (ATCC14458)	0.5**
	Staphylococcus epidermidis (ATCC1228)	0.5**
	Staphylococcus epidermidis (6epi)	Resistant
Annona salzmannii (Costa et al. 2011)	Escherichia coli (ATCC 10538)	1**
	Pseudomonas aeruginosa (ATCC27853)	Resistant
	Staphylococcus aureus (ATCC6538)	Resistant
	Staphylococcus aureus (ATCC14458)	0.5**
	Staphylococcus epidermidis (ATCC1228)	0.5**
	Staphylococcus epidermidis (6epi)	1**
Annona squamosa (Elhawary et al. 2013; Mohammed et al. 2016)	Bacillus cereus[2]	31*
	Bacillus subtilis[2]	26*
	Bacillus subtilis (ATCC 6051)[1]	16*
	Escherichia coli (ATCC 11775)[1]	15*
	Pseudomonas aeruginosa[2]	22*
	Pseudomonas aeruginosa (ATCC 10145)[1]	16*
	Staphylococcus aureus[2]	21*
	Staphylococcus aureus (ATCC 12600)[1]	17*
Annona vepretorum (Costa et al. 2012)	Escherichia coli (ATCC 10538)	Resistant
	Pseudomonas aeruginosa (ATCC27853)	Resistant
	Staphylococcus aureus (ATCC6538)	Resistant
	Staphylococcus aureus (ATCC14458)	0.5**
	Staphylococcus epidermidis (ATCC1228)	0.5**
	Staphylococcus epidermidis (6epi)	Resistant
Annona cherimola x Annona squamosa Hybrid (Mohammed et al. 2016)	Bacillus cereus	19*
	Bacillus subtilis	22*
	Pseudomonas aeruginosa	25*
	Staphylococcus aureus	20*

[1] first reference; [2] second reference; * diameter of growth inhibition (mm); ** minimum inhibitory concentration in mg/mL^{-1}.

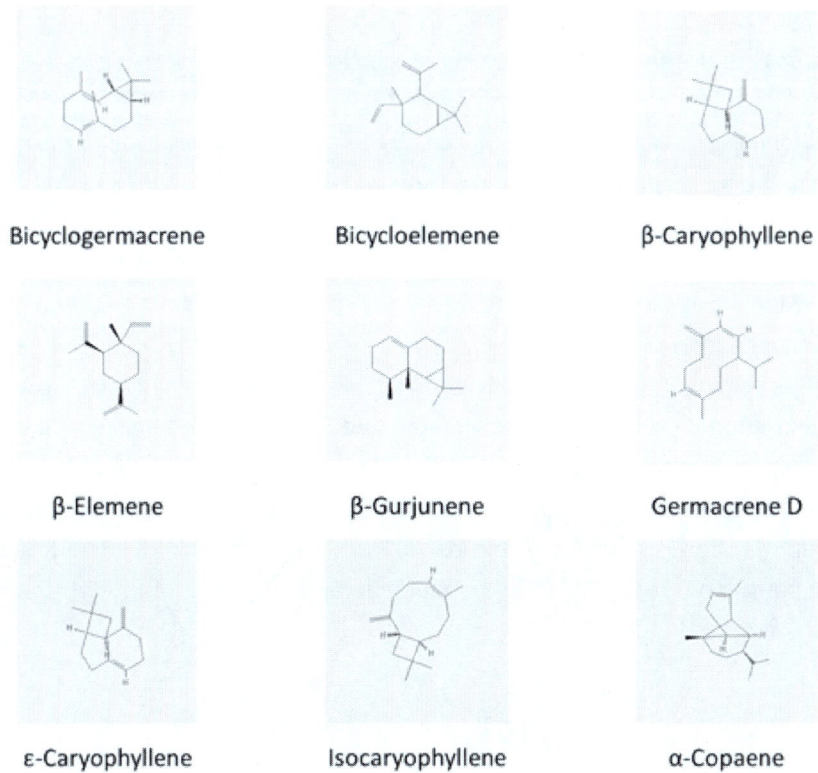

Figure 2. Major volatile compounds found in essential oils from species of Annona.

Bacillus and *Staphylococcus* strains were the most tested against *Annona* species essential oils, but *Pseudomonas aeruginosa* was the one tested with all species. Two species, *Annona cherimola* and *Annona squamosa*, were reported twice. However, both studies used different strains (Elhawary et al. 2013; Mohammed et al. 2016). The average of antibacterial results through minimum inhibitory concentration was 0.5 mg/mL (Elhawary et al. 2013; Mohammed et al. 2016). With the diameter of growth inhibition method, *Annona* species reduced about 20 mm (Costa et al. 2012, 2011). The essential oil from a hybrid specie from *Annona cherimola* and *Annona squamosa* was also tested; but its results did not differ from tests of both species separately, even though the volatile profile was different from both species (Mohammed et al. 2016).

The activity of all species was somewhat similar, but *Annona squamosa* and *Annona cherimola* were the strongest (Elhawary et al. 2013; Mohammed et al. 2016). Strains of *Escherichia coli* (ATCC 10538), *Pseudomonas aeruginosa* (ATCC27853), *Staphylococcus aureus* (ATCC6538), and *Staphylococcus epidermidis* (6epi) were resistant to *Annona pickelli* (Costa et al. 2011), and *Annona vepretorum* (Costa et al. 2012) essential oils. Moreover, *Pseudomonas aeruginosa* (ATCC27853) and Staphylococcus aureus (ATCC6538) were resistant to Annona salzmannii (Costa et al. 2011).

The *Annona* species that were inefficient to some strains had lower concentrations of β-Caryophyllene and ε-Caryophyllene, volatile compounds related to greater antibiotic effect (Chavan, Shinde, and Nirmal 2006; Foddai et al. 2019; Deans et al. 1995; Zhang et al. 2017). For those strains that these species did combat, minimum inhibitory concentrations were considered moderate to high, which demonstrate lack of effectiveness (Costa et al. 2011, 2012).

CONCLUSION

Plenty species of *Annona* still need tests with their essential oil. Essential oil of *Annona squamosa* and *Annona cherimola* were the most potent, while *Annona pickelli*, *Annona salzmannii* and *Annona vepretorum* were the weakest. Hybrid of *Annona cherimola* x *Annona squamosal* presented different composition of volatile compounds, showing possibilities for new hybrids in order to obtain greater concentrations of specific constituents of essential oil. The lack of studies and divergence in methodologies reduced clinical applicability of *Annona* species essential oils as antibiotic, although they seem very promising.

REFERENCES

Biba, V. S., P. W. Jeba Malar, and P. Remani. 2013. "Antibacterial Activity of Annona Squamosa Seed Extract." *International Journal of Pharmacy and Technology* 5 (3): 5651-59.

Biba, V. S., A. Amily, S. Sangeetha, and P. Remani. 2014. "Anticancer, Antioxidant and Antimicrobial Activity of *Annonaceae* Family." *World Journal of Pharmacy and Pharmaceutical Sciences* 3 (3): 1595-1604.

Bicas, Juliano Lemos, Gustavo Molina, Ana Paula Dionísio, Francisco Fábio Cavalcante Barros, Roger Wagner, Mário Roberto Maróstica, and Gláucia Maria Pastore. 2011. "Volatile Constituents of Exotic Fruits from Brazil." *Food Research International* 44 (7): 1843-55. https://doi.org/10.1016/j.foodres.2011.01.012.

Chavan, M. J., D. B. Shinde, and S. A. Nirmal. 2006. "Major Volatile Constituents of Annona Squamosa L. Bark." *Natural Product Research* 20 (8): 754-57. https://doi.org/10.1080/14786410500138823.

Chen, Ya Yun, Chen Xiao Peng, Yan Hu, Chen Bu, Shu Chen Guo, Xiang Li, Yong Chen, and Jian Wei Chen. 2017. "Studies on Chemical Constituents and Anti-Hepatoma Effects of Essential Oil from Annona Squamosa L. Pericarps." *Natural Product Research* 31 (11): 1305-8. https://doi.org/10.1080/14786419.2016.1233411.

Costa, Emmanoel Vilaça et al. 2011. "Chemical Composition and Antioxidant, Antimicrobial, and Larvicidal Activities of the Essential Oils of Annona Salzmannii and A. Pickelii (*Annonaceae*)." *Natural Product Communications* 6 (6): 1934578X1100600. https://doi.org/10.1177/1934578x1100600636.

Costa, Emmanoel Vilaça, Lívia Macedo Dutra, Paulo Cesar de Lima Nogueira, Valéria Regina de Souza Moraes, Marcos José Salvador, Luis Henrique Gonzaga Ribeiro, and Fernanda Ramos Gadelha. 2012. "Essential Oil from the Leaves of Annona Vepretorum: Chemical Composition and Bioactivity." *Natural Product Communications* 7 (2): 1934578X1200700. https://doi.org/10.1177/1934578x1200700240.

Deans, S. G., R. C. Noble, R. Hiltunen, W. Wuryani, and L. G. Pénzes. 1995. "Antimicrobial and Antioxidant Properties of Syzygium Aromaticum (L.) Merr. & Perry: Impact upon Bacteria, Fungi and Fatty Acid Levels in Ageing Mice." *Flavour and Fragrance Journal* 10 (5): 323-28. https://doi.org/10.1002/ffj.2730100507.

El-Chaghaby, Ghadir A., Abeer F. Ahmad, and Eman S. Ramis. 2014. "Evaluation of the Antioxidant and Antibacterial Properties of Various Solvents Extracts of Annona Squamosa L. Leaves." *Arabian Journal of Chemistry* 7 (2): 227-33. https://doi.org/10.1016/j.arabjc.2011.06.019.

Elhawary, Seham S., Mona E. El Tantawy, Mohamed a. Rabeh, and Noha E. Fawaz. 2013. "DNA Fingerprinting, Chemical Composition, Antitumor and Antimicrobial Activities of the Essential Oils and Extractives of Four Annona Species from Egypt." *Journal of Natural Sciences Research* 3 (13): 59-69. https://iiste.org/Journals/index.php/JNSR/article/view/9067.

Evangelho, Jarine Amaral do, Guilherme da Silva Dannenberg, Barbara Biduski, Shanise Lisie Mello el Halal, Dianini Hüttner Kringel, Marcia Arocha Gularte, Angela Maria Fiorentini, and Elessandra da Rosa Zavareze. 2019. "Antibacterial Activity, Optical, Mechanical, and Barrier Properties of Corn Starch Films Containing Orange Essential Oil." *Carbohydrate Polymers* 222 (April): 114981. https://doi.org/10.1016/j.carbpol.2019.114981.

Foddai, Marzia, Mauro Marchetti, Alessandro Ruggero, Claudia Juliano, and Marianna Usai. 2019. "Evaluation of Chemical Composition and Anti-Inflammatory, Antioxidant, Antibacterial Activity of Essential Oil of Sardinian Santolina Corsica Jord. & Fourr." *Saudi Journal of Biological Sciences* 26 (5): 930-37. https://doi.org/10.1016/j.sjbs.2018.08.001.

Hu, Wei, Changzhu Li, Jinming Dai, Haiying Cui, and Lin Lin. 2019. "Antibacterial Activity and Mechanism of Litsea Cubeba Essential Oil against Methicillin-Resistant Staphylococcus Aureus (MRSA)." *Industrial Crops and Products* 130 (November 2018): 34-41. https://doi.org/10.1016/j.indcrop.2018.12.078.

Javed, Aamir, Kumari Annu, M N Khan, and S K Medam. 2013. "Evaluation of the Combinational Antimicrobial Effect of Annona Squamosa and Phoenix Dactylifera Seeds Methanolic Extract on Standard Microbial Strains." *International Research Journal of Biological Sciences* 2 (5): 68-73. http://www.isca.in/IJBS/Archive/v2/

i5/11.ISCA-IRJBS-2013-052.pdf%5Cnhttp://ovidsp.ovid.com/ovidweb.cgi?T=JS&CSC=Y&NEWS=N&PAGE=fulltext&D=cagh&AN=2013 3202416%5Cnhttp://oxfordsfx.hosted.exlibrisgroup.com/oxford?sid=OVID:caghdb&id=pmid:&id=doi:&issn=2278-320.

Lin, Lin, Xuefang Mao, Yanhui Sun, Govindan Rajivgandhi, and Haiying Cui. 2019. "Antibacterial Properties of Nanofibers Containing Chrysanthemum Essential Oil and Their Application as Beef Packaging." *International Journal of Food Microbiology* 292 (May 2018): 21-30. https://doi.org/10.1016/j.ijfoodmicro.2018.12.007.

Mohammed, M A, S E El-Gengaihi, A M A Enein, E M Hassan, O K Ahmed, and M S Asker. 2016. "Chemical Constituents and Antimicrobial Activity of Different Annona Species Cultivated in Egypt." *Journal of Chemical and Pharmaceutical Research* 8 (4): 261-71.

Neethu, Simon K, R Santhoshkumar, and Neethu S Kumar. 2016. "Phytochemical Analysis and Antimicrobial Activities of Annona Squamosa (L) Leaf Extracts." *Journal of Pharmacognosy and Phytochemistry* 5 (4): 128-31. http://www.phytojournal.com/archives/2016/vol5issue4/PartB/5-3-73-646.pdf.

Padhi, L P, S K Panda, S N Satapathy, and S K Dutta. 2011. "*In Vitro* Evaluation of Antibacterial Potential of Annona Squamosa L. and Annona Reticulata L. from Similipal Biosphere Reserve, Orissa, India." *International Journal of Agricultural Technology* 7 (1): 133-42.

Pinto, Nícolas de C.C., Lara M. Campos, Anna Carolina S. Evangelista, Ari S.O. Lemos, Thiago P. Silva, Rossana C.N. Melo, Caroline C. de Lourenço, et al. 2017. "Antimicrobial Annona Muricata L. (Soursop) Extract Targets the Cell Membranes of Gram-Positive and Gram-Negative Bacteria." *Industrial Crops and Products* 107 (December 2016): 332-40. https://doi.org/10.1016/j.indcrop.2017.05.054.

Rabelo, Suzana Vieira, Jullyana de Sousa Siqueira Quintans, Emmanoel Vilaça Costa, Jackson Roberto Guedes da Silva Almeida, and Lucindo José Quintans. 2015. *Annona Species (Annonaceae) Oils. Essential Oils in Food Preservation, Flavor and Safety*. Elsevier Inc. https://doi.org/10.1016/B978-0-12-416641-7.00024-9.

Rinaldi, Maria V.N., Ingrit E.C. Díaz, Ivana B. Suffredini, and Paulo R.H. Moreno. 2017. "Alkaloids and Biological Activity of Beribá (Annona Hypoglauca)." *Brazilian Journal of Pharmacognosy* 27 (1): 77-83. https://doi.org/10.1016/j.bjp.2016.08.006.

Ruddaraju, Lakshmi Kalyani, Panduranga Naga Vijay Kumar Pallela, S. V. N. Pammi, Veerabhadra Swamy Padavala, and Venkata Ramana Murthy Kolapalli. 2019. "Synergetic Antibacterial and Anti-carcinogenic Effects of Annona Squamosa Leaf Extract Mediated Silver Nano Particles." *Materials Science in Semiconductor Processing* 100 (February): 301-9. https://doi.org/10.1016/j.mssp.2019.05.007.

Salazar, Gerson Javier Torres, Jéssica Pereira de Sousa, Cicera Norma Fernandes Lima, Izabel Cristina Santiago Lemos, Ana Raquel Pereira da Silva, Thiago Sampaio de Freitas, Henrique Douglas Melo Coutinho, Luiz Everson da Silva, Wanderlei do Amaral, and Cícero Deschamps. 2018. "Phytochemical Characterization of the Baccharis Dracunculifolia DC (Asteraceae) Essential Oil and Antibacterial Activity Evaluation." *Industrial Crops and Products* 122 (March): 591-95. https://doi.org/10.1016/j.indcrop.2018.06.052.

Wang, Fei, Fuyao Wei, Chunxiao Song, Bin Jiang, Shangyi Tian, Jingwen Yi, Chunlei Yu, et al. 2017. "Dodartia Orientalis L. Essential Oil Exerts Antibacterial Activity by Mechanisms of Disrupting Cell Structure and Resisting Biofilm." *Industrial Crops and Products* 109 (August): 358-66. https://doi.org/10.1016/j.indcrop.2017.08.058.

Zhang, Yi, Yue Wang, Xiaojing Zhu, Ping Cao, Shaomin Wei, and Yanhua Lu. 2017. "Antibacterial and Antibiofilm Activities of Eugenol from Essential Oil of Syzygium Aromaticum (L.) Merr. & L. M. Perry (Clove) Leaf against Periodontal Pathogen Porphyromonas Gingivalis." *Microbial Pathogenesis* 113 (September): 396-402. https://doi.org/10.1016/j.micpath.2017.10.054.

Chapter 8

ANTIMICROBIAL ACTIVITY OF ESSENTIAL OIL FROM *XYLOPIA FRUTESCENS* AUBL (*ANNONACEAE*)

Amanda Virginia Barbos[1,]*,
Bárbara de Azevedo Ramos[1],
Milena Martins Correia da Silva[2],
Claudio Augusto Gomes da Camara[2],
Rafael Artur Cavalcanti Queiroz de Sá[1],
Francisco Henrique da Silva[1],
Maria Tereza dos Santos Correia[1]
and Marcia Vanusa da Silva[1]

[1]Department of Biochemistry, Federal University of Pernambuco,
Recife, Pernambuco, Brazil
[2]Department of Chemistry, Federal Rural University of Pernambuco,
Recife, Pernambuco, Brazil

---

* Corresponding Author's Email: amandavirginia88@gmail.com.

ABSTRACT

Plants with medicinal properties have been used by humans since ancient times as an alternative resource of treatment for various diseases. *Xylopia frutescens* popularly known as "embira", is used in folk medicine and produces essential oil from its secondary metabolism. The use of essential oils in phytotherapy is associated to various pharmacological actions and are extensively studied mainly for their antimicrobial activity. Thus, this study aims to chemically characterize the components present in the essential oil of leaves of *X. frutescens* and evaluation of its antimicrobial activity. Twenty five components have been identified in the essential oil. The minimal inhibitory concentration (MIC) of the essential oil was at 4 mg / mL for all tested strains of *Staphylococcus aureus*. In conclusion, this is the first report on analysis of volatile compounds and antimicrobial activity from *X. frutescens* essential oil of leaves extracted from Atlantic Forest in the Pernambuco, Brazil. The essential oil of *X. frutescens* showed satisfactory MIC results for all *S. aureus* strains, indicating a promising source of biomolecules for development of new antibacterial drugs.

INTRODUCTION

Plants with medicinal properties are used by humans from ancient times to the present time and are used as therapeutic resources for the treatment of various diseases (Carneiro et al. 2014). A large portion of emerging countries populations uses traditional medicine as initial alternative or as the only source of healthcare treatments (Wannes; Tounsi; Marzouk 2017). In Brazil, the increasing of herbal medicines usage may be related to two important factors. Due to scientific advances, it was allowed a safer and efficient use of medicine herbs. The other reason was the search for therapies with low side effects (Bruning; Mosegui; Vianna 2012). In this scenario, plants act as a source of alternative drugs in the treatment of many healthcare problems and being widely used as raw material for synthesis of bioactive compounds (Al-Rubaye; Hameed; Kadhim 2017).

Natural products derived from plants are already known for their pharmacological activities to combat diseases such as cancer and diabetes (De Corte 2016; Alam et al. 2018) and its antimicrobial and antiparasitic properties (Singh and Mahajan 2013). For many years, plants have been used as therapeutic agents and many of the commercial medicines available come from natural sources, especially plants (Atanasov et al. 2015).

Brazil stands out for its immense biodiversity. A large portion of its native plants still unexplored and distributed in their biomes (Martins et al. 2016).

The Atlantic Forest is one of the Brazilian biomes that is present on coastal region. It is one of the most important tropical forests in the world (Pereira 2009). There are a large number of plant species (mostly endemic), which are promising sources of biomolecules.

Annonaceae family is present in the Atlantic Forest bioma. Its representatives are well studied for their pharmacological effects such as: analgesic, antispasmodic, antioxidant action, among others (Nishiyama et al. 2006; Paula et al. 2014). In this family, *Xylopia frutescens* is a tree popularly known as "embira" or "embira vermelha". It is used in folk medicines to treat rheumatism, inflammation and diarrhea (Sena Filho et al. 2008; Souza et al. 2015).

In addition, pharmacological studies attributed other activities to *X. frutescens* as anti-trypanosome and antitumor (Silva et al. 2013; Ferraz et al. 2013). These activities are related to the components present in extracts, which can be obtained from many parts of the plant or volatile compounds present in leaves.

Among natural compounds synthesized by plants, essential oils are increasingly becoming the focus of several researches (Fonseca et al. 2015; Miranda et al. 2016; Liang et al. 2017; Sarma et al. 2019). There is estimated that three thousand essential oils have already been described, of which three hundred are destined for fragrance market (Martins et al. 2016) and others may be a source of new bioactive compounds.

The use of essential oils in phytotherapy is associated to many pharmacological activities attributed to their constituents and are extensively studied mainly for their antimicrobial activities (Millezi et al. 2014). Once all constituents of essential oils have been elucidated and their pharmacological potential confirmed, they can be used as natural sources to substitute or being used in association with synthetic drugs.

Antimicrobial activity is the starting point for several studies including the development of new drugs. In recent times, many infections caused by resistant microorganisms cannot be treated with commonly prescribed antimicrobials in clinic. Even last-resort antibiotics are losing their effect (Frieri; Kumar; Boutin 2017). Thus, plants with medicinal properties are getting space and increasingly being used in treatment of diseases in order to alleviate collateral effects brought by prolonged use of antibiotics (Freire et al. 2014).

Consequently, the pharmaceutical industry has been investing in bioprospecting researches. They are looking for the potential of plants to provide potential candidate molecules for new drugs (Montanari; Bolzani 2001).

Based on these considerations, this study aims to characterize chemically the components present in the essential oil of *Xylopia frutescens* Aubl leaves from the Pernambuco Atlantic Forest. Besides, to evaluate its antimicrobial activity against multiresistant pathogens.

MATERIALS AND METHODS

Plant Material

Leaves of *X. frutescens* were collected in June 2015 at Usina São José - PE. After collection, the plant samples were properly identified and stored for later use.

ESSENTIAL OIL EXTRACTION

The leaves of *X. frutescens* were crushed and subjected to extraction of the essential oils by the hydrodistillation technique using a Clevenger type apparatus for 4h. The yield was calculated by the dry weight of the plant material. The essential oil *X. frutescens* (XFEO) was stored at -20°C and protected from light.

CHEMICAL COMPOSITION

Analysis Qualitative

The analysis was carried using GC/MS (Varian 220-MS IT GC system) with a mass selective detector, mass spectrometer in EI 70 eV with a scan interval of 0.5s and fragments from 40 to 550Da. fitted with the same column and temperature program as that for the GC experiments, with the following parameters: carrier gas = helium; flow rate = 1mL/min; split mode (1:30); injected volume = 1µL of diluted solution (1:100) of oil in n-hexane.

Analysis Quantitative

The analysis was carried out using a Hewlett-Packard 5890 Series II GC apparatus equipped with a flame ionization detector (FID) and a non-polar DB-5 fused silica capillary column (30m x 0.25mm x 0.25mm µm film thickness) (J & W Scientific).

The oven temperature was programmed from 60 to 240°C at a rate 3°C/min for integration purposes. Injector and detector temperatures were 260°C.

Hydrogen was used as the carrier gas at a flow rate of 1L/min and 30 p.s.i. inlet pressure in split mode (1:30).

The injection volume was 0.5µL of diluted solution (1/100) of oil in n-hexane.

The amount of each compound was calculated from GC peak areas in the order of DB-5 column elution and expressed as relative percentage of the total area of the chromatograms.

Identification of Components

Identification of the components was based on GC retention indices with reference to a homologous series of C8-C40 n-alkanes calculated using the Van den Dool and Kratz equation (Van den Dool; Kratz 1963) and by computer matching against the mass spectral library of the GC/MS data system (NIST 98 and WILEY) and co-injection with authentic standards as well as other published mass spectra (Adams 2007).

ANTIMICROBIAL ACTIVITY EVALUATION

Bacterial Samples and Growing Conditions

Clinical isolates of *Staphylococcus aureus* from the Collection of Microorganisms of the Department of Antibiotics of the Federal University of Pernambuco (UFPEDA) from different sites of infection were used. The standard strain 02 UFPEDA, was used as reference control in all experiments performed (CLSI 2014). All cultures were kept in mineral oil and grown in Petri dishes containing Mueller-Hinton Agar medium at 37°C.

Susceptibility Test

The *S. aureus* isolates were tested for antimicrobial susceptibility of different mechanisms of action: Ampicillin (AMP) 10μg, Oxacillin (OXA) 1μg, Ciprofloxacin (CIP) 5μg, Amicacin (AMI) 30μg, Gentamicin (GEN) 10μg, Clidamycin (CLI) 2μg, Chlorafenicol (CLO) 30μg and Tetracycline (TET) 30μg. The experiments were performed in triplicate by antibiogram using the disk diffusion method (Kirby-Bauer) according to Clinical and Laboratory Standards Institute (CLSI 2014). The plates were incubated for 24h at 37°C. Inhibition zone diameters were measured and compared with the standard table for antibiotic susceptibility tests.

Minimum Inhibitory Concentration (MIC) and Minimum Bactericidal Concentration (MBC)

The antimicrobial evaluation of XFEO was performed in an experimental serial microdilution model, as described by CLSI (2014) in 96-well plates. The bacterial suspension was adjusted to Optical Density (OD600) between 0.150 to 0.200 and then plates containing culture medium and samples (microorganism and XFEO) were incubated at 37°C for 24 hours. Microplate reading was performed by spectrophotometer at a wavelength of 600nm at times 0h and 24h. MIC was determined as the lowest concentration capable of inhibiting 90% of bacterial growth. MBC was performed under the same conditions as MIC and determined as the lowest concentration of the compound for which no viable bacteria was observed. The antibiotic tetracycline was used as control.

Statistical Analysis

All analyzes were performed in triplicate, calculated as means and standard deviation. P values < 0.05 were accepted as statistically significant compared to the control group.

Results and Discussion

Chemical Characterization

The hydrodistillation process of *X. frutescens* leaves yielded in average 0.65% of essential oil.

Table 1. Chemical composition of the essential oil of *X. frutescens* (XFEO)

Compound	RI [a]	RI [b]	Area (%)
(E)-β-ocimene	1039	1044	0.46
α-cubebene	1342	1345	0.70
Longipinene	1347	1350	0.58
β-patchoulene	1372	1379	0.75
β-cubebene	1383	1387	0.84
β-caryophyllene	1414	1417	13.62
β-humulene	1431	1436	1.39
α-guaiene	1433	1437	0.34
aromadendrene	1436	1439	0.40
α-humulene	1448	1454	5.50
Allo aromadendrene	1454	1558	6.91
trans-cadina-1(6),4-diene	1472	1475	0.45
γ-muurolene	1473	1478	0.90
Germacrene D	1482	1484	1.09
Viridiflorene	1490	1496	27.19
Bicyclogermacrene	1496	1500	30.91
α-muurolene	1501	1500	0.46
γ-cadinene	1508	1513	0.23
β-curcumene	1511	1514	0.26
β-sesquiphellandrene	1517	1521	0.92
δ-cadinene	1519	1522	1.96
γ-vetivene	1526	1531	0.91
10-epi-cubebol	1529	1533	0.93
α-cadinene	1531	1537	0.50
Germacrene	1550	1559	0.20
Total			98.40
Monoterpenes			0.46
Sesquiterpenes			97.94
Total			98.40

[a] Retention index calculated from retention times in relation to those of a series of n-alkanes on a 30m DB-5 capillary column. [b] Linear retention index from the literature.

Twenty-five components of essential oil of *X. frutescens* (Table 1) were identified, representing 98.40% of the chemical composition of the oil.

The oil was rich in sesquiterpenes (97.94%). Bicyclogermacrene (30.91%) was the major constituent, followed by Viridiflorene (27.19%), β-caryophyllene (13.62%), Allo aromadendrene (6.91%) and α-humulene (5.50%).

Results found in this study are in agreement with data reported in the literature for essential oil of *X. frutescens* from different localities of the Northeast region (Souza et al. 2015; Nascimento et al. 2017) and Southeast Brazil (Ferraz et al. 2013). The presence of bicyclogermacrene, γ-muurolene, spathulenol, germancrene D, and some other compounds identified in the XFEO is reported in essential oils of other species of *Annonaceae* family. The analysis of chemical composition of essential oil from *Annona foetida* leaves showed that its major component was Bicyclogermacrene (35.12%). Other compounds such as: α-humulene (2.22%), Aromadendrene (1.81%), Allo aromadendrene (1.51%), and Spathulenol (1.11%) were also present, but in small concentrations (Costa et al, 2009).

Da Silva et al. (2013), found out that the main components of essential oil from *X. laevigata* leaves were Germacrene D (27.0%), bicyclogermacrene (12.8%), (E) -caryophyllene (8.6%), γ-muurolene (8.6%), δ-cadinene (6, 8%), and Germacrene B (6.0%). This shows that some of these compounds can be found in *Xylopia genus*.

Antimicrobial Activities

Antimicrobial resistance corresponds to relative insusceptibility that a particular microorganism has to a specific treatment under certain conditions (Carneiro et al. 2007). However, when there is a change in susceptibility that causes an ineffective action of antibiotic against a particular microorganism, this organism is known as resistant. This resistance is due to the physiological and biochemical factors of each particular organism (Kümmerer 2004).

For the susceptibility test, 9 strains of *S. aureus* were tested to determine their susceptibility profile to different antibiotics (Table 2).

The tested antibiotics represent important drug classes for the treatment of bacterial infections. Most isolates showed resistance to more than one antibiotic class. Ampicillin, Oxacillin, and Ciprofloxacine were the classes with more resistance. Uncontrolled use of antibiotics is certainly one of the factors that contributes to increased resistance (Guimarães; Momesso; Pupo 2010).

In fact, the number of multiresistant bacteria has been grown at an alarming rate causing serious damage to health systems and resulting in high mortality rates (Bessimbaye et al. 2015). This contributes to search new drugs. Antimicrobial substances of natural origin have been increased as an alternative source. Essential oils are sources of numerous molecules and have been extensively studied due to their antimicrobial properties (Tajkarimi et al. 2010; Lucena et al. 2015; Miranda et al. 2016; Yuan et al. 2018).

Minimum Inhibitory Concentration (MIC) is a widely antimicrobial method used with microorganism. It defines the lowest concentration of a substance that inhibits the growth of organisms after a certain incubation period (24 hours or more for anaerobes) (Ferrer et al. 2016). The XFEO for all strains tested had a MIC of 4mg/mL. MIC results were moderate, but were able to inhibit the growth of multiresistant *S. aureus*.

Some studies relate the major constituents present in essential oils as responsible for their biological activities (Ziaei et al. 2011; Govindarajan and Benelli, 2016).

Mendes et al. (2017), related the antimicrobial action exerted by essential oil *X. sericea* against Gram-negative and Gram-positive bacteria, especially *S. aureus,* to sesquiterpene spathulenol. Costa et al. (2009), analyzed the essential oil of *A. foetida*. It was observed that this essential oil is rich in sesquiterpernes and showed important antimicrobial activity. The authors attributed this activity to the joint presence of Bicyclogermacrene and (E) – caryophyllene.

Table 2. Resistance profile and antimicrobial activity of the essential oil *X. frutescens* (XFEO)

Strains	Clinical source	Antimicrobials (resistance)	XFEO (mg/mL)		Tetracycline MIC (mg/mL)
			MIC	MBC	
02	Type Strain	Sensitive	4	> 4	0.1
679	Surgery wound	AMP, OXA	4	> 4	0.0125
683	Purulent exudate	AMP, OXA, CIP, AMI, GEN	4	> 4	0.5
699	Catheter	AMP, OXA	4	> 4	> 1
700	Ulcer secretion	AMP, OXA, CIP, AMI, GEN, CLI, CLO, TET	4	> 4	0.25
705	Surgery wound	CIP	4	> 4	<0.03125
709	Purulent exudate	AMP, OXA, CIP, AMI, GEN, CLI, CLO	4	> 4	0.0625
731	Surgery wound	AMP, CLI, CLO	4	> 4	<0.03125
733	Bone	AMP, CLI, CIP, CLO	4	> 4	<0.03125

MBC: Minimum Bactericidal Concentration; MIC: Minimal Inhibitory Concentration; XFEO: X. frutescens Essential Oil.

Thus, the antimicrobial activity of XFEO may be related to the presence of Bicyclogermacrene, which is also reported for antifungal activity (Silva et al. 2007). Regarding MBC, the highest tested concentration of the essential oil (4mg/mL), was not bactericidal against multiresistant *S. aureus* strains.

Essential oils are used for antimicrobial testing due to their efficient bacteriostatic as well as microbicidal action. Sobeh et al. (2016), analyzed the essential oil *Eugenia uniflora*, they found that even at the highest tested oil concentration (10mg/mL), it was not effective against *S. aureus* (ATCC). Similar results were observed by Martins et al. (2016), for the essential oil of *Aeollanthus suaveolens* against *Escherichia coli*, *Salmonella* sp., and *S. aureus*. The authors reported that the concentration of 100mg/mL was not effective for *S. aureus*. These studies differ from the results obtained in the present work.

CONCLUSION

This study was a first report on the analysis of components and antimicrobial activity of the essential oil of *X. frutescens* from the Pernambuco Atlantic Forest. Based on these results, we could show that XFEO presented antimicrobial activity against multiresistant *S. aureus* strains. In addtion, it can be considerate as a promising source for the development of new antibacterial drugs.

REFERENCES

Adams, Robert. P. 2007. *"Identification of Essential Oil Components by Gas Chromatography/Quadrupole Mass Espectroscopy"*. 4th ed. Allured Publishing Corporation: Carol Stream, p804. ISBN 978-1-932633-11-4.

Alam, Fahmida et al. 2018. "Updates on Managing Type 2 Diabetes Mellitus with Natural Products: Towards Antidiabetic Drug Development". *Curr. Med. Chem.*, 25(39): 5395 - 543. Doi:10.2174/0929867323666160813222436.

Al-Rubaye, Abeer Fauzi et al. 2017. "A Review: Uses of Gas Chromatography-Mass Spectrometry (GC-MS) Technique for Analysis of Bioactive Natural Compounds of Some Plants". *International Journal of Toxicological and Pharmacological Research*, 9(1): 81 - 85.

Atanasov, Atanas G. et al. 2015. "Discovery and resupply of pharmacologically active plant-derived natural products: A review". *Biotechnology Advances*, 33(8): 1582 - 1614. Doi: 10.1016/ j.biotechadv.2015.08.001.

Bessimbaye, Nadlaou et al. 2015. "Prevalence Multi-Resistant Bacteria in Hospital N'djamena, Chad". *Chemo. Open Access*, 4(4):170. Doi:10.41 72/2167-7700.1000170.

Bruning, Maria Cecilia Ribeiro et al. 2012. "The use of phytotherapy and medicinal plants in primary health care units in the cities of Cascavel and Foz do Iguaçu - Paraná: the viewpoint of health professionals". *Ciência and Saúde Coletiva*, 17(10):2675 - 2685.

Carneiro, D. O. et al. 2007. "Profile of antimicrobial resistance in bacterial populations recovered from different Nile tilapia (*Oreochromis niloticus*) culture systems". *Arq. Bras. Med. Vet. Zootec.*, 59(4): 869 - 876. Doi: 10.1590/S0102-09352007000400008.

Carneiro, Fernanda Melo et al. 2014. "Trends of studies for medicinal plants in Brazil". *Revista Sapiência: sociedade, saberes e práticas educacionais* [*Sapiência Magazine: society, knowledge and educational practices*], 3(2): 44 - 75. ISSN 2238-3565.

CLSI - "Perfomance Standards for Antimicrobial Susceptibility Testing; Twenty-Fourth Informational Supplement". 2014. CLSI document M100-S24. Wayne, PA. Clinical and Laboratory Standards Institute.

Costa, Vilaça Emmanoel et al. 2009. "Antimicrobial and antileishmanial activity of essential oil from the leaves of *Annona foetida* (*Annonaceae*)". *Quim. Nova.*, 32(1): 78 - 81. Doi:10.1590/S0100-40422009000100015.

Da Silva, Thanany et al. 2013. "Chemical composition and anti-Trypanosoma cruzi activity of essential oils obtained from leaves of *Xylopia frutescens* and *X. laevigata* (*Annonaceae*)". *Nat. Prod. Commun.*, 8(3):403 - 406. Doi: 10.1177/1934578X1300800332.

De Corte, Bart 2016. "Underexplored Opportunities for Natural Products in Drug Discovery". *Journal of Medicinal Chemistry*, 59 (20): 9295 - 9304. DOI: 10.1021/acs.jmedchem.6b00473.

Ferraz, Rosana P. C. et al. 2013. "Antitumour properties of the leaf essential oil of *Xylopia frutescens* Aubl. (*Annonaceae*)". *Food Chemistry*, 141(1):196 - 200. Doi: 10.1016/j.foodchem.2013.02.114.

Ferrer, Manuel et al. 2016. "Antibiotic use and microbiome function". *Biochemical Pharmacology*, 134: 114 - 126. Doi: 10.1016/ j.bcp.2016. 09.007.

Fonseca, M. C. M. et al. 2015. "Potential of essential oils from medicinal plants to control plant pathogens". *Rev. bras. plantas med.*, 17(1) Doi:10.1590/1983-084X/12_170.

Freire, I. C. M. et al. 2014. *"Antibacterial Activity of Essential Oils against Strains of Streptococcus and Staphylococcus"*. *Rev. Bras. Pl. Med.*, 16(2):372 - 377. Doi: 10.1590/1983-084X/12_053.

Frieri, Marianne et al. 2017. "*Antibiotic resistance"*. *Journal of Infection and Public Health*, 10(4): 369 - 378. Doi: 10.1016/j.jiph.2016.08.007.

Govindarajan, Marimuthu and Benelli, Giovanni 2016. "Eco-friendly larvicides from Indian plants: Effectiveness of lavandulyl acetate and bicyclogermacrene on malaria, dengue and Japanese encephalitis mosquito vectors". *Ecotoxicology and Environmental Safety*, 133: 395 - 402. Doi: 10.1016/j.ecoenv.2016.07.035.

Guimarães, Denise Oliveira et al. 2010. "Antibiotics: therapeutic importance and perspectives for the discovery and development of new agents". *Quim. Nova*, 33(3): 667 - 679. Doi: 10.1590/S0100-40422010 000300035.

Kümmerer, Klaus 2004. "Resistance in the environment". *Journal of Antimicrobial Chemotherapy*, 54 (2): 311 - 320. Doi: 10.1093/jac/dkh 325.

Liang, Jun-yu et al. 2017. "Bioactivities and Chemical Constituents of Essential Oil Extracted from *Artemisia anethoides* Against Two Stored Product Insects". *Journal of Oleo Science*, 66(1):71 - 76. Doi: 10.5650/ jos.ess16080.

Lucena, Bruno F. F. et al. 2015. "Evaluation of Antibacterial Activity of Aminoglycosides and Modulating the Essential Oil of *Cymbopogon citratus* (DC.) Stapf". *Acta Biol. Colomb.*, 20(1):39 - 45. Doi:10.1544 6/abc.v20n1.41673.

Martins, Rosany Lopes et al. 2016. "Chemical Composition, an Antioxidant, Cytotoxic and Microbiological Activity of the Essential

Oil from the Leaves of *Aeollanthus suaveolens* Mart. ex Spreng". *PLoS One*, 11(12):e0166684. Doi: 10.1371/journal.pone.0166684.

Mendes, Renata F. et al. 2017. "The essencial oil from the fruits of the Brazilian spice *Xylopia sericea* A. St.-Hil. Presentes expressive invitro antibaterial and antioxidant activity". *Journal of Pharmacy and Pharmacology*, 54(12): 3093 - 3102. Doi: 10.1111/jphp.12698.

Millezi, Alessandra Farias et al. 2014. "Chemical characterization and antibacterial activity of essential oils from medicinal and condiment plants against *Staphylococcus aureus* and *Escherichia coli*". *Rev. Bras. Pl. Med.*, 16(1): 18 - 24. Doi: 10.1590/S1516-05722014000100003.

Miranda, Cíntia Alvarenga Santos Fraga et al. 2016. "Essential oils from leaves of various species: antioxidant and antibacterial properties on growth in pathogenic species". *Revista Ciência Agronômica*, 47(1): 213 - 220, 2016. DOI: 10.5935/1806-6690.20160025.

Montanari, Carlos Alberto and Bolzani, Vanderlan da Silva 2001. "Drug design based on natural products". *Quim. Nova*, 24(1):105 - 111. DOI: 10.1590/S0100-40422001000100018.

Nascimento, Ana M. D. et al. 2017. "Repellency and larvicidal activity of essential oils from *Xylopia laevigata, Xylopia frutescens, Lippia pendunculosa*, and their individual compounds against *Aedes aegypti* Linnaeus". *Neotrop. Entomol.*, 46(2): 223 - 230. Doi: 10.1007/s13744-016-0457-z.

Nishiyama, Yumi et al. 2006. "Secondary and tertiary isoquinoline alkaloids from Xylopiaparviflora". *Phytochemistry*, 67: 2671 - 2675. Doi: 10.1016/j.phytochem.2006.07.011.

Paula, C. S. et al. 2014. "Antioxidant activity and preliminary toxicity of the extracts and fractions obtained from the leaves and stem bark of *Dasyphyllum tomentosum* (Spreng.) Cabrera". *Rev. Bras. Pl. Med.*, 16(2): 189 - 195. Doi: 10.1590/S1516-05722014000200004.

Pereira, Anísio Baptista 2009. "Atlantic rain forest: a geographic approach". *Nucleus*, 6(1). Doi: 10.3738/1982.2278.152.

Sarma, Riju et al. 2019. "Combinations of Plant Essential Oil Based Terpene Compounds as Larvicidal and Adulticidal Agent against

Aedes aegypti (Diptera: Culicidae)". *Scientific Reports*, 19(1):9471. Doi: 10.1038/s41598-019-45908-3.

Sena Filho, José G. et al. 2008. "Preliminary Phytochemical Profile and Characterization of the Extract from the Fruits of *Xylopia frutescens* Aubl. (*Annonaceae*)". *Journal of Essential Oil Research*, 20(6):536 - 538. Doi: 10.1080/10412905.2008.9700082.

Silva, Luciana da et al. 2007. "Bicyclogermacrene, resveratrol and fungitoxic activity on leaves extracts of *Cissus verticillata* L. Nicolson and Jarvis (Vitaceae)". *Brazilian Journal of Pharmacognosy*, 17(3):361 - 367. Doi: 10.1590/S0102-695X2007000300010.

Silva, Thanany Brasil da et al. 2013. "Chemical composition and anti-trypanossoma cruzi activity of essential oils obtained from leaves of *Xylopia frutescens* and *X. laevigata* (*Annonaceae*)". *Nat. Prod. Commun.*, 8:403 - 406. Doi: 10.1177/1934578X1300800332.

Singh, Inder Pal and Mahajan, Shivani. 2013. "Berberine and its derivatives: a patent review (2009 - 2012)". *Expert Opin. Ther. Patents*, 23(2) 215 - 231. Doi: 10.1517/13543776.2013.746314.

Sobeh, Mansour et al. 2016. "Chemical Profiling of the Essential Oils of *Syzygium aqueum*, *Syzygium samarangense* and *Eugenia uniflora* and Their Discrimination Using Chemometric Analysis". *Chemistry and Biodiversity*, 13(11): 1537 - 1550. Doi: 10.1002/cbdv.201600089.

Souza, Iara Leão Luna de et al. 2015. "Essential oil from *Xylopia frutescens* Aubl. Reduces cytosolic calcium level songuine apigileum: mechanis munderlying its spasmolytic potential". *BMC Complementary and Alternative Medicine*, 15. Doi: 10.1186/s12906-015-0849-3.

Tajkarimi, Mehrdad M. et al. 2010. "Antimicrobial herb and spice compounds in food". *Food Control*, 21:1199 - 1218. Doi: 10.1016/ j.fo odcont.2010.02.003.

Van den Dool, H. and Kratz, P. D. 1963. "A generalization of the retention index system including linear temperature programmed gas-liquid partition chromatography". *J. Chromatogr.*, *A*. 11:463 - 471. Doi: 10.1016/S0021-9673(01)80947-X.

Wannes, Wisseem Aidi et al. 2017. "A review of Tunisian medicinal plants with anticancer activity". *Journal of Complementary and Integrative Medicine,* 15(1):1 - 14. Doi: 10.1515/jcim-2017-0052.

Yuan, Juanjuan et al. 2018. "Composition and antimicrobial activity of the essential oil from the branches of *Jacaranda cuspidifolia* Mart. growing in Sichuan, China". *Natural Product Research,* 32(12): 1451 - 1454 Doi: 10.1080/14786419.2017.1346644.

Ziaei, Akram et al. 2011. "Identification of Spathulenol in *Salvia mirzayanii* and the immunomodulatory effects". *Phytoterapy Research,* 25: 557 - 562. Doi: 10.1002/ptr.3289.

ABOUT THE EDITORS

Fernanda Granja de Oliveira graduated in Pharmaceutical Sciences from the Federal University of Paraíba, master's and doctorate in Biotechnology from the Federal University of Sergipe. Participates in a research group that investigates the potential of Flora da Caatinga, in addition to the molecular responses of plants to biotic and abiotic stresses. He has experience in the field of macromolecule biochemistry with an emphasis on essential oils and molecules of secondary metabolism of plants, acting mainly on the following themes: purification, characterization and biological applications of secondary metabolites and essential oils of plants.

Bruno Oliveira de Veras holds a degree in Biological Sciences from the Federal University of Paraíba, a master's degree in Biotechnology from the Federal University of Pernambuco and a doctor in Tropical Medicine from the Federal University of Pernambuco. Participates in a research group that investigates the potential of Flora da Caatinga, in addition to the molecular responses of plants to biotic and abiotic stresses. He has experience in the field of macromolecule biochemistry with an emphasis on essential oils and molecules of secondary metabolism of plants, acting mainly on the following themes: purification, characterization and biological applications of secondary metabolites and essential oils of plants.

Yago Queiroz dos Santos graduated in Biological Sciences (2014) from the Federal University of Paraíba and master's degree in Biochemistry and Molecular Biology (2016) from the Federal University of Rio Grande do Norte - UFRN. Currently develops research at the Chemical and Bioactive Protein Function Laboratory (LQFPB), of the Department. Biochemistry Department at UFRN and at the Microbiology Laboratory of the Institute of Tropical Medicine (IMT-UFRN). He has experience in the areas of Biochemistry and Genetics of Microorganisms, with special focus on the isolation and characterization of enzymes, molecular identification of microorganisms and determination of the antibacterial activity of bioactive molecules. secondary plant, acting mainly on the following themes: purification, characterization and biological applications of secondary metabolites and essential oils of plants.

Jackson Roberto Guedes da Silva Almeida holds a degree in Pharmacy from the Federal University of Paraíba (2001) and a Master's degree (2004) and a Doctorate (2006) in Natural and Bioactive Synthetic Products from the Postgraduate Program in Natural and Bioactive Synthetic Products from the Federal University of Paraíba. He did Post-Doctorate (2013) at the Faculty of Pharmaceutical Sciences at USP - Ribeirão Preto. He is an Associate Professor of the Pharmacy Course at the Federal University of Vale do São Francisco (UNIVASF), where he coordinates research and extension projects in the areas of medicinal plants and phytotherapy. He held the position of Coordinator of the Graduate Program in Natural Resources in the Semi-Arid Region of UNIVASF (2011-2015). He is an advisor professor at the Graduate Program in Biosciences at UNIVASF. Advisor professor at the Biotechnology Graduate Program of the Northeast Biotechnology Network (RENORBIO). Advisor Professor of the Post-Graduate Program in Biotechnology at the State University of Feira de Santana (UEFS). He has experience in the field of macromolecule biochemistry with an emphasis on essential oils and molecules of secondary metabolism of plants, acting mainly on the following themes: purification, characterization and biological applications of secondary metabolites and essential oils of plants.

Maria Tereza dos Santos Correia has a degree in Industrial Chemistry from the Federal University of Pernambuco (1980), a master's degree in Biochemistry from the Federal University of Pernambuco (1989) and a doctorate in Biological Sciences (Molecular Biology) from the Federal University of São Paulo (1995). He is currently Full Professor at the Federal University of Pernambuco. She served as coordinator and vice-coordinator of the Graduate Program in Biological Sciences (PPGCB-UFPE), deputy coordinator of the Master's Degree in Biochemistry at UFPE, Chief and sub-chief of the Department of Biochemistry; at different times. She works as an undergraduate in the Biomedicine Course (UFPE) and as a Permanent Professor at PPGCB, the Graduate Program in Biochemistry and Physiology (PPGBF) and the Graduate Program in Biotechnology (RENORBIO). He has experience in the biochemistry of macromolecules with an emphasis on proteins, essential oils and molecules of secondary metabolism of plants, acting mainly on the following themes: purification, characterization and biological applications of proteins (plants and animals), secondary metabolites and essential oils of plants.

Márcia Vanusa da Silva holds a degree in Agronomic Engineering from the Federal University of Pernambuco (1997), a master's degree in Genetics and Plant Breeding from the Federal University of Lavras (1989) and a PhD in Cellular and Molecular Biology from the Federal University of São Paulo (1995). Associate Professor, Department of Biochemistry, Federal University of Pernambuco. Participates in a research group that investigates the potential of Flora da Caatinga, in addition to the molecular responses of plants to biotic and abiotic stresses. He has experience in the field of macromolecule biochemistry with an emphasis on essential oils and molecules of secondary metabolism of plants, acting mainly on the following themes: purification, characterization and biological applications of secondary metabolites and essential oils of plants.

INDEX

A

Annona, vii, xii, 139, 142, 143, 144, 145, 146, 147, 148, 149, 150, 151, 152, 153, 154, 163, 167
Annona L., vii, 143
Annonaceae, vii, xii, 137, 139, 143, 144, 145, 151, 153, 155, 157, 163, 167, 170
antibacterial, v, vii, x, xi, xii, xiii, 4, 6, 9, 12, 15, 17, 18, 20, 21, 22, 28, 30, 33, 35, 36, 39, 40, 41, 45, 47, 51, 54, 55, 57, 58, 59, 60, 63, 66, 70, 74, 76, 77, 78, 86, 95, 117, 118, 119, 120, 121, 122, 123, 124, 125, 126, 127, 128, 129, 133, 136, 137, 138, 139, 140, 142, 143, 145, 147, 149, 150, 152, 153, 154, 156, 166, 168, 169
Antibiofilm activity, 7, 13, 18
antibiotic, x, 15, 16, 45, 52, 54, 55, 56, 57, 59, 60, 61, 62, 63, 64, 67, 74, 77, 118, 132, 142, 145, 150, 160, 161, 163, 164, 168
antifungal activity, 18, 35, 78, 79, 80, 89, 95, 123, 125, 128, 165
antimicrobial, v, vi, vii, ix, x, xi, xii, 1, 2, 3, 4, 5, 6, 9, 10, 11, 14, 15, 16, 17, 18, 19, 20, 22, 30, 31, 32, 34, 39, 42, 46, 47, 48, 54, 55, 61, 62, 63, 64, 65, 66, 69, 70, 71, 74, 75, 76, 77, 78, 79, 81, 83, 84, 85, 86, 87, 88, 89, 91, 93, 94, 95, 96, 97, 98, 99, 100, 103, 109, 116, 117, 118, 119, 120, 121, 122, 123, 124, 125, 126, 127, 128, 130, 131, 132, 137, 139, 140, 141, 144, 145, 146, 147, 151, 152, 153, 155, 156, 157, 158, 160, 161, 163, 164, 165, 166, 167, 168, 170, 171
antimicrobial activity, vi, vii, ix, x, xi, xii, 2, 3, 4, 5, 9, 11, 14, 15, 17, 19, 22, 30, 31, 42, 54, 62, 70, 71, 75, 76, 77, 78, 81, 84, 85, 93, 95, 97, 100, 103, 109, 117, 118, 120, 121, 122, 123, 124, 125, 126, 127, 130, 132, 137, 140, 141, 144, 145, 146, 147, 151, 153, 155, 156, 158, 160, 164, 165, 166, 171
antimicrobial potential, xi, 4, 10, 76, 94, 96, 99, 117, 131, 132, 145
aromatherapy, v, x, 12, 17, 21, 22, 23, 24, 25, 26, 27, 35, 39, 40, 41, 42, 43

B

bacteria, ix, x, xi, xii, 1, 7, 12, 13, 15, 31, 33, 34, 38, 42, 45, 46, 47, 48, 49, 51, 52, 56, 58, 61, 62, 63, 64, 65, 66, 70, 71, 74,

Index

75, 76, 77, 79, 84, 85, 94, 99, 116, 117, 120, 130, 132, 136, 139, 140, 141, 142, 144, 145, 151, 153, 161, 164, 167
bacterial resistance, vi, x, 11, 45, 47, 51, 55, 56, 58, 60, 66, 125
bicyclogermacrene, 100, 101, 102, 111, 113, 146, 162, 163, 165, 168, 170
biochemical prospection, xii, 94, 95
biofilm, 5, 6, 7, 8, 11, 13, 14, 15, 16, 17, 20, 63, 132, 138, 145, 154

C

caatinga plant, vi, 93
chemical composition, 2, 10, 17, 18, 19, 34, 41, 53, 72, 84, 118, 120, 121, 122, 123, 124, 125, 128, 139, 140, 151, 152, 159, 163, 168
clove, 60, 85, 130, 131, 132, 135, 137, 138, 140, 141, 142, 154
compositions, 3, 15, 16, 17, 18, 20, 24, 25, 31, 34, 39, 40, 41, 42, 43, 46, 53, 54, 70, 72, 73, 74, 85, 86, 88, 89, 117, 118, 119, 120, 121, 122, 123, 125, 126, 127, 128, 142, 146, 150, 162, 167, 170, 171

E

efflux pump, x, 45, 46, 47, 51, 56, 57, 58, 60, 64, 65
emulsion, 9, 10, 13
escherichia coli, 106
essential oils, v, vi, vii, ix, x, xi, xii, 1, 2, 3, 4, 5, 6, 7, 8, 9, 10, 11, 12, 13, 14, 15, 16, 17, 18, 19, 20, 21, 22, 23, 24, 25, 26, 27, 28, 29, 30, 31, 32, 34, 35, 36, 37, 38, 39, 40, 41, 42, 43, 45, 46, 53, 54, 55, 56, 57, 58, 59, 60, 61, 62, 63, 64, 65,66, 67, 69, 70, 71, 72, 73, 74, 78, 79, 80, 81, 82, 83, 84, 86, 87, 88, 89, 90, 91, 93, 94, 95, 96, 97, 98, 100, 103, 109, 117, 118, 119, 120, 121, 122, 123, 124, 125, 126, 127, 128, 129, 130, 131, 132, 133, 134, 135, 137, 138, 139, 140, 141, 142, 143, 144, 145, 146, 147, 149, 150, 151, 152, 153, 154, 155, 156, 157, 158, 161, 162, 163, 164, 165, 166, 167, 168, 169, 170, 171
eugenol, xii, 4, 36, 56, 60, 63, 101, 113, 114, 130, 131, 132, 133, 134, 135, 136, 137, 138, 141, 142, 154

I

interaction, 6, 7, 67, 137

M

mechanisms to antibiotics, 47
medicinal plants, ix, 2, 12, 13, 63, 119, 121, 127, 167, 168, 171
microencapsulation, 9, 18, 20
microorganisms, ix, xi, 1, 3, 5, 6, 7, 9, 10, 31, 33, 46, 53, 54, 70, 71, 75, 77, 81, 84, 94, 100, 101, 102, 103, 104, 105, 106, 107, 108, 109, 110, 111, 112, 113, 114, 115, 116, 117, 120, 124, 132, 145, 158, 160, 161, 163, 164

O

oil components, 42

P

phenolic compounds, 4, 14, 54
potent antimicrobial, vi, 69

R

Resistance of B-Lactamases, 54

resistance to antibiotics, ix, xi, 1, 2, 5, 45, 46

S

species, vi, vii, x, xi, xii, 7, 18, 22, 25, 29, 30, 32, 33, 34, 35, 40, 41, 42, 43, 46, 49, 57, 58, 60, 63, 70, 71, 72, 73, 78, 79, 80, 81, 86, 93, 94, 95, 96, 97, 98, 99, 100, 101, 102, 103, 104, 105, 106, 107, 108, 109, 110, 111, 112, 113, 114, 115, 116, 117, 120, 121, 123, 127, 128, 131, 132, 133, 134, 135, 138, 143, 144, 145, 146, 147, 148, 149, 150, 152, 153, 157, 163, 169

Staphylococcus aureus, xi, xiii, 7, 10, 11, 12, 13, 14, 18, 19, 20, 49, 57, 64, 66, 70, 74, 75, 77, 82, 83, 87, 95, 99, 100, 101, 102, 103, 104, 105, 106, 107, 108, 109, 110, 111, 112, 113, 114, 115, 119, 120, 134, 135, 147, 148, 150, 156, 160, 169

Syzygium aromaticum, vii, xii, 36, 129, 130, 132, 133, 135, 137

Syzygium Aromaticum, 137, 138, 139, 140, 141, 142, 151, 154

T

tea tree, vi, xi, 14, 40, 42, 55, 61, 69, 70, 71, 74, 75, 78, 80, 81, 82, 83, 84, 85, 86, 87, 88, 89, 90, 119

terpenes, xi, 3, 5, 53, 69, 70, 74, 95

V

volatile compounds, xiii, 4, 7, 53, 132, 136, 145, 149, 150, 156, 157

Related Nova Publications

THORACIC LYMPHADENOPATHY

EDITOR: Vikas Pathak, MD

SERIES: New Developments in Medical Research

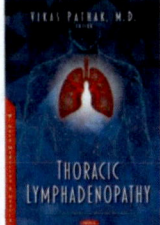

BOOK DESCRIPTION: In this book, we have tried to simplify the lymph node stations based on the latest IASLC guidelines, done a very comprehensive review about mediastinal and hilar lymphadenopathy in different disease states and provided the pathway to diagnosis.

SOFTCOVER ISBN: 978-1-53616-700-9
RETAIL PRICE: $82

A CLOSER LOOK AT WOUND INFECTIONS AND HEALING

EDITOR: Joseph E. Keel

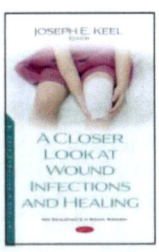

SERIES: New Developments in Medical Research

BOOK DESCRIPTION: Wound healing is a complex cascade of events that led to reconstruct a damaged tissue with cellular and biological mechanisms. *A Closer Look at Wound Infections and Healing* first reviews the treatments mentioned in traditional Iranian medicine sources for various wounds.

SOFTCOVER ISBN: 978-1-53616-816-7
RETAIL PRICE: $95

To see a complete list of Nova publications, please visit our website at www.novapublishers.com

Related Nova Publications

Toward Precision Assessment and Psychotherapy: Understanding Individual Differences through Neurobiology, Genetics, and Epigenetics

Author: Thomas G. Arizmendi, PhD

Series: New Developments in Medical Research

Book Description: "Toward Precision Assessment and Psychotherapy: Understanding Individual Differences through Neurobiology, Genetics, and Epigenetics" provides a transformative approach to the understanding of mental health in the 21st century.

Hardcover ISBN: 978-1-53616-536-4
Retail Price: $230

Natural Anti-Aging Plants and Delay of Senescence

Editor: Noboru Motohashi, PhD

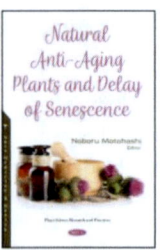

Series: New Developments in Medical Research

Book Description: Worldwide, human life is headed for longevity. Based on these ideas, this book will focus on plant ingredients and plants that can be expected to maintain health until this longevity.

Softcover ISBN: 978-1-53616-282-0
Retail Price: $95

To see a complete list of Nova publications, please visit our website at www.novapublishers.com